continued...

"A wonderful portrayal of the conflicting emotions of doubt and certainty in the scientists' minds as they tell themselves repeatedly that Einstein has to be right. This is a very high-stakes game and the players are, in many ways, gamblers. Their strategies, strengths, and weaknesses are engagingly constructed and artfully told."

—David DeVorkin, curator, History of
Astronomy and the Space Sciences, National Air
and Space Museum, Smithsonian Institution

"Marcia Bartusiak brings to vivid life the people and personalities, the challenges and controversies surrounding the hunt for gravitational waves. This is a sneak preview of what gravitational-wave astronomy will be all about."

—Clifford M. Will, author of *Was Einstein Right?*

"Bartusiak shows the reader the real world of science, where hypotheses are championed or challenged by people equipped not only with powerful minds but also courageous vision, fragile egos and ample eccentricities."   —*Minneapolis Star Tribune*

"A beautiful history of gravitational-wave physics."

—Alan Lightman, author of *Einstein's Dreams*

ALSO BY MARCIA BARTUSIAK

Thursday's Universe

Through a Universe Darkly

A Positron Named Priscilla
as co-author

MARCIA BARTUSIAK

Einstein's Unfinished Symphony

Listening to
the Sounds of
Space-Time

BERKLEY BOOKS, NEW YORK

A Berkley Book
Published by The Berkley Publishing Group
A division of Penguin Putnam Inc.
375 Hudson Street
New York, New York 10014

PRINTING HISTORY
Joseph Henry Press hardcover edition / December 2000
Berkley trade paperback edition / February 2003

Library of Congress Cataloging-in-Publication Data

Bartusiak, Marcia, 1950–
    Einstein's unfinished symphony : listening to the sounds of space-time / Marcia
Bartusiak.
        p. cm.
    Originally published: Washington, D.C. : Joseph Henry Press, c2000.
    Includes bibliographical references and index.
    ISBN 0-425-18620-2 (pbk.)
    1. Space and time. 2. General relativity (Physics) I. Title.

QC173.59.S65 B39 2003
530.11—dc21

                                                                              2002028388

PRINTED IN THE UNITED STATES OF AMERICA

10   9   8   7   6   5   4   3   2   1

For Steve

Contents

Acknowledgments   xi

Prelude   1

Space in G-Flat   11

The Maestro Enters   23

Starlight Waltz   47

Pas de Deux   67

Bars and Measures   87

Dissonant Chords   117

A Little Light Music   149

Variations on a Theme   175

The Music of the Spheres   191

Finale   213

Coda   231

Bibliography   233

Index   241

Acknowledgments

I was formally introduced to the science of gravitational waves nearly two decades ago. I was on assignment for *Science 85*, a magazine (now regrettably gone) that thrived on covering the cutting edge of science. Stanford University in California then housed the most advanced instrument yet constructed in the quest to detect gravitational waves—a hulking, supercooled bar of metal that resided in a cavernous room on the campus. This five-ton aluminum bar was the principal subject of my article on the continuing search for the elusive ripples in space-time. The best such a detector could hope to see, though, was a supernova explosion in our galaxy, which occurs only a couple of times each century.

While on the west coast, almost as an afterthought, I decided to drop by Caltech to learn about another detection scheme that promised to see more sources in the long run but was described as being in its "infancy" compared to the bars: laser interferometry. Caltech housed a prototype on campus and during my visit I was shown a crude drawing of the full-blown observatory they hoped to construct one day in collaboration with MIT. The picture displayed two giant tubes that sprawled for miles over an imaginary desert plain. At the time, with federal budget cuts so prominent, I doubted I would ever see such an instrument built in my lifetime.

To my surprise and delight I was wrong. I use the word delight for good reason. At the time I began to write on astronomy and astrophysics, the various means of observing the universe's electromagnetic radiations were fairly well established. Astronomers had gotten a good start at examining the cosmos over the entire spectrum, from radio waves to gamma rays. I figured that the era had passed when a science writer could chronicle the advent of an entirely new astronomy, where the heavens were a blank slate ready to be filled in. But gravity wave astronomy, I came to realize, now offered me that opportunity.

For showing me the way, I must initially thank Peter Saulson, whom I first met in 1988 as I was writing a follow-up story on the field's progress. Over the intervening years, as Peter moved from MIT to Syracuse University, he kept me abreast of the advancing technology and planted the notion to expand my magazine coverage into a book. His encouragement initiated the project; his sage advice and guidance followed me through to its completion. He and his wife Sarah have become cherished friends in the process.

When not involved with either interviewing or traveling—from Pasadena to Pisa—I could usually be found at the Science Library of the Massachusetts Institute of Technology by the banks of the Charles River. I would like to thank the many archivists and librarians there, who so patiently answered my questions and pointed me in the right direction toward the stacks.

By the end of my research, I had interviewed over 50 scientists and engineers, either in person, over the phone, or via e-mail communications. My appreciation is extended to those who kindly agreed to review selected sections of my book, catching my errors and providing additional insights. They are John Armstrong, Peter Bender, GariLynn

Billingsley, Philip Chapman, Karsten Danzmann, Sam Finn, Ron Drever, William Folkner, Robert Forward, Adalberto Giazotto, William Hamilton, Russell Hulse, Richard Isaacson, Albert Lazzarini, Fred Raab, Roland Schilling, Irwin Shapiro, Joseph Taylor, Kip Thorne, Tony Tyson, Stan Whitcomb, Clifford Will, and Michael Zucker. (Any errors that do remain are entirely of my own doing.) I am particularly grateful to Rainer Weiss, who in the midst of deadline pressures during LIGO's commissioning always welcomed me and found the time to answer my questions. Robert Naeye, a favorite editor of mine at *Astronomy* magazine, also provided a keen editorial eye.

Thanks must also be given to those physicists and researchers who provided an essential historic perspective. They include John Wheeler, who took me on a delightful walk around Princeton University, where I got to see many of Einstein's old haunts, and Joseph Weber, who was a gracious host during my visit to the University of Maryland. For informative tours of the various detector sites, I thank Jennifer Logan at the Caltech prototype, Carlo Bradaschia at VIRGO, Cecil Franklin at the Louisiana LIGO, and Otto Matherny at the Hanford LIGO. At Caltech, Donna Tomlinson, administrator extraordinaire, offered invaluable assistance in arranging my visit to the LIGO headquarters. I am equally grateful to the investigators who provided either key resource material or insightful discussions of their work. They are Barry Barish, Biplab Bhawal, Rolf Bork, Jordan Camp, Mark Coles, Douglas Cook, Robert Eisenstein, Jay Heefner, Jim Hough, Vicky Kalogera, Ken Libbrecht, Phil Lindquist, Walid Majid, Dale Ouimette, E. Sterl Phinney, Janeen Hazel Romie, R. John Sandeman, Gary Sanders, David Shoemaker, R. Tucker Stebbins, Serap Tilav, Wim van Amersfoort, Robbie Vogt, and Hiro Yamamoto.

Throughout the course of this project, friends and colleagues kept my spirits high by staying in touch, even as I would disappear for weeks on end. Their interest in my progress served as an incentive to keep moving forward. For this I thank Elizabeth and Goetz Eaton, Elizabeth Maggio and Ike Ghozeil, Tara and Paul McCabe, Suzanne Szescila and Jed Roberts, Fred Weber and Smita Srinivas, Linda and Steve Wohler, L. Cole Smith, Ellen Ruppel Shell, and Dale Worley.

Meanwhile, my goddaughter Skye McCole Bartusiak brought me smiles from afar with her varied entertainments. I am also grateful for the unwavering support of my mother, my brother Chet and his family, my sister Jane Bailey and her family, as well as Clifford, Eunice, and Bob Lowe, my husband's family. A special thank-you is extended to Duffy, my favorite neighborhood dog, who would energetically bark at my door to make sure I remembered it was important to get up from the computer, go outside, and walk around to smell the roses from time to time.

The reader would not be viewing this work at all, though, if it were not for the singular persistence of Stephen Mautner, executive editor of the Joseph Henry Press, whose powers of persuasion are formidable. He never faltered in his faith in this project and patiently waited nearly two years for me to sign on. (Thank you, Margaret Geller, for introducing us.) I have now discovered what a joy it is to work with a publishing house where effective communication of science is a priority and not just a sideline. Steve makes it all possible. Others at Joseph Henry that I must thank for their enthusiastic support behind the scenes are Barbara Kline Pope, Robin Pinnel, and Ann Merchant. Christine Hauser was an immense help in tracking down elusive photos, while copy editor Barbara Bodling skillfully honed my manuscript to a fine sheen.

Lastly, I must acknowledge my husband, Steve Lowe, whose love, encouragement, and support throughout these last two years allowed my ideas to reach fruition. More than that, I came to depend on his superb editorial judgment on matters scientific. Thank you, Steve, for being there.

Prelude

Ah, but a man's reach should exceed his grasp,
Or what's a heaven for?
— Robert Browning, *Andrea del Sarto*, 1855

To arrive at the astronomical observatory of the twenty-first century, you must journey through America's Old South—Baton Rouge, Louisiana, to be precise. Coming off a plane, visitors are first greeted by the pungent smells wafting from a nearby petroleum refinery, a major state industry. Continuing along Interstate 10, which partly follows the winding path of the Mississippi River on its way to New Orleans and the Gulf of Mexico, billboards hawk snuff, Killer Joe's seafood, and Louisiana mud painting. East of the state capital in plantation country, the land turns flat. The roads are arrow-straight, like the unswerving grids on a sheet of graph paper. The regularity is relieved only by sinuous chains of trees, green rivers of foliage laden with Spanish moss, that occasionally meander through the farmland.

Louisiana has nearly 14 million acres of forests. Longbed trucks piled high with cut lumber are a familiar sight on the highways. Many of these trucks originate about 25 miles east of Baton Rouge, in the parish of Livingston, where a vast pine reserve resides. A few miles past a deserted feed store, north along Highway 63, a modest sign announces the presence of the Laser Interferometer Gravitational-Wave Observatory, operated by the California Institute of Technology (Caltech) and the Massachusetts Institute of Technology (MIT). Those in the know simply call it LIGO (pronounced lie-go). Nothing can be seen from the highway, though. You must first turn onto an asphalt road and slowly drive past young-growth forests, some sections newly cut. Longhorn cattle lazily graze along the roadway. After a mile's ride the observatory at last comes into view. It resembles a multistory warehouse, silver-gray in color, with blue and white trim. Observatories are not a common sight in Louisiana, but establishing new technologies is a state tradition. In 1811 Edward Livingston, who would later become a U.S. senator and the parish's namesake, helped Robert Fulton establish the first commercial steamboat operation on the Mississippi River.

The vast room where LIGO's key instruments are mounted soars upward for more than 30 feet. The cross-shaped hall resembles an aircraft hangar or perhaps the transept and nave of a modern-day basilica. On two ends of the cross, at right angles to each other, are large round ports. Attached to each of these two openings is a long tube, which extends out into the countryside for 2½ miles. Each four feet wide, the two tubes resemble oil pipelines and are roomy enough for a crablike walk should the need arise. To accommodate the pipes, the pine forest has been leveled. A huge, raw L has been carved into the woody terrain to make room for these lengthy metal arms. One long swathe stretches to the southeast, the other to the southwest. Alongside each arm a roadway has been constructed about 8 feet above the Louisiana floodplain. The dirt for the roadway was dug up right on site, creating two water-filled canals, each parallel to a tube. An alligator, fed donuts by the construction crew, even adopted one of the borrow pits as his home. The pipes cannot be seen directly. Concrete covers, 6 inches thick, protect them from the wind and rain, as well as any stray bullet that might pass by during hunting season. A hit could be devastating, for the pipes are as empty of air as the vacuum of space.

"Here's our 2½-mile-long hole in the atmosphere," says Mark Coles, sweeping his hand outward with pride. A big and friendly man, Coles recently moved from Caltech to direct the Livingston observatory. He took to the Cajun lifestyle with ease, even coming up with a bit of French for the observatory's slogan, displayed on T-shirts sold in the reception area: *Laissez les bonnes ondes rouler!* (Let the good waves roll!) The signals they seek are waves of gravitational radiation—gravity waves in common parlance.*

Electromagnetic waves, be they visible light, radio, or infrared waves, are released by individual atoms and electrons and generally reveal a celestial object's physical condition—how hot it is, how old it is, or what it is made of. Gravity waves will convey different information. They will tell us about the overall motions of massive celestial objects. They are literally quakes in space-time that will emanate from the most violent events the universe has to offer—a once blazing sun burning out and going supernova, the dizzying spins of neutron stars, the cagey dance of two black holes pirouetting around one another, approaching closer and closer until they merge. Gravity waves will tell us how large amounts of matter move, twirl, and collide throughout the universe. Eventually, this new method of examining the cosmos may even record the remnant rumble of the first nanosecond of creation, what remains of the awesome space-time jolt of the Big Bang itself. At the dedication of the Livingston observatory, Rita Colwell, director of the National Science Foundation, noted that they were "breaking a bottle of champagne on the bow of a figurative galleon that will take us back farther in time than we have ever been." So compelling is this information that scientists have been pushing the envelope of technology to detect these subtle tremors.

---

*In physics, the term *gravity wave* has long been used officially to refer to another phenomenon—an atmospheric disturbance in which a wave originates from the relative buoyancy of gases of different density. For simplicity's sake, I use the term here as a synonym for *gravitational wave*, as researchers themselves often do when speaking casually. To quote Shakespeare, the term's succinctness comes "trippingly on the tongue . . . that may give it smoothness."

Inside LIGO's main hall the ambiance is almost reverential, akin to the response one might feel inside a darkened telescope dome, the mirror aimed for the star-dappled heavens. But this astronomical venture is vastly different. Here there are no windows to spy on the universe. A gravitational wave observatory is firmly planted on cosmos firma. It is glued to space-time, awaiting its faint rumbles, vibrations first predicted by Einstein more than 80 years ago.

"The worship of Einstein, it's the only reason we're here, if you want to know the truth," says Rainer Weiss of MIT. "There was this incredible genius in our midst, in our own lifetime. The average person knows that he was an important guy. If you go to Congress and tell them you're going to try to show that Heisenberg's uncertainty principle is not quite right, you run into blank stares. But if you say you're measuring something that's proving or disproving Einstein's theory, then all sorts of doors open. There's a mystique."

Albert Einstein indeed stands like a giant amid the pantheon of scientific figures of the twentieth century. His ideas unleashed a revolution whose changes are still being felt into the new century. What he did was to take our commonplace notions of space and time and completely alter their definitions. The evolution of physics in many ways traces our changing conceptions of space and time. And each adjustment in our worldview—our cosmic frame of reference—brought with it an accompanying change in the physics of handling that new framework. It is the start of every physicist's musings: an attempt to track an object's motion in space and time. These intangibles are given names, such as "miles" and "seconds," and ever since Sir Isaac Newton analyzed that falling apple, scientists believed they finally understood the true meanings of those words. But they didn't. Einstein stepped in to shatter that confidence. With his general theory of relativity, he showed us that matter, space, and time are not separate entities but rather eternally linked, producing the force known as gravity. That is why Einstein initiated a revolution. He taught us that space and time are not mere definitions useful for measurements. Instead, they are joined together as an object known as space-time, whose geometric shape is determined by the matter around it. According to general relativity, massive bodies, such as stars, dimple the space-time around them (much the way a bowling ball sitting on a trampoline would create a depression). Planets and comets are then attracted to the star simply because they are following the curved space-time highway carved out by the star. Nature as we like to think of it, which flows so nicely according to Newton's laws, is actually a special case, where energies and velocities are low. Many who fall under the spell of Einstein feel like Alice in Wonderland: the universe begins to look "curiouser and curiouser." With his more universal law, Einstein introduced us to a world that is counterintuitive, a place where lengths can contract, time can speed up or slow down, and matter can disappear in a wink down a space-time well. He made space and time palpable.

To the practiced eye, Einstein's equations stand as the quintessence of mathematical beauty. When it was first introduced in 1915, general relativity was hailed as a momentous conceptual achievement. But for a long time it was thought to have little practical importance. While the theory was embraced—and Einstein was recognized for his genius—the evidence for it was largely aesthetic because the tests possible in the early twentieth century were few: a tiny shift observed in the planet Mercury's orbit and evidence of starlight bending around the Sun due to the indentation of space around the orb's enormous mass. There was a perfectly good reason that experimental tests lagged behind theorists' contemplation of relativity's effects. In either describing the motion of a ball falling to the ground or sending a spacecraft to the Moon, Newton's laws of gravity are still adequate. General relativity is far more subtle, best observed when gravitational fields attain monstrous strengths, making its effects more prominent. But in Einstein's day the universe was deemed more tame than that. After the first few tests were completed—elevating Einstein to the pinnacle of celebrity—general relativity became largely a theoretical curiosity, admired by all but exiled into the hinterland of physics. "General relativists," says Clifford Will, a relativist himself, "had the reputation of residing in intellectual ivory towers, confining themselves to abstruse calculations of formidable complexity."

But like some theoretical Rip Van Winkle, general relativity gradually revived after decades of neglect, especially as astronomers came to discover a host of intriguing celestial objects, such as pulsars, quasars, and black holes, that could be understood only through the physics of general relativity. Neutron stars, gravitational lenses, inflationary universes—all must be studied with Einstein's vision in mind. At the same time, advancing technology offered physicists new and ingenious means to test relativity's quirky and subtle effects with unprecedented accuracy—and not just in the laboratory. Using planetary probes, radio telescopes, and spaceborne clocks as their tools, investigators have checked Einstein's hypotheses with uncanny degrees of precision. The entire solar system now serves as general relativity's laboratory. In the words of general relativity experts Charles Misner, Kip Thorne, and John Wheeler, from their book *Gravitation*, a veritable bible for workers in the field, "General relativity is no longer a theorist's Paradise and an experimentalist's Hell." For nearly a century the theory has held up to every experimental test, which is not only a triumph for Einstein—a celebration of his accomplishment—but also a further example of nature's ability to inspire awe and fascination in the way its rules follow such a precise mathematical blueprint. Furthermore, there are practical reasons why the theory is no longer relegated to the ivory tower. General relativity now has a real impact on our everyday lives. To properly operate the satellites of the Global Positioning System, used regularly by hikers, mariners, and soldiers to keep track of their locations, requires corrections of Einsteinian precision, revisions unaccounted for by Newton's cruder take on gravity. And astronomers who electronically link radio telescopes on several continents—creating a telescope as big as the world for their observations—also demand such accuracy.

And yet the story of general relativity remains incomplete. A secret still resides within general relativity, a major prediction that awaits *direct* confirmation: gravity waves. To understand this phenomenon, imagine one of the most violent events the universe has to offer—a supernova, the spectacular explosion of a star. More than 160,000 years ago, at a time when woolly mammoths were walking the Asian plains, a brilliant blue-white star known as Sanduleak −69° 202 exploded in the Large Magellanic Cloud, a prominent celestial landmark in the southern sky. Not until the winter of 1987 did a wave of particles and electromagnetic radiation shot from that dying star reach the shores of Earth. And when it arrived, an arsenal of observatories around the world focused their instruments on the flickers of light and energy that represented the star's long-ago death throes. It was the first time astronomers were able to observe a supernova in our local galactic neighborhood since the invention of the telescope.

But Einstein's theory suggests that, when Sanduleak −69° 202 blew, it also sent out waves of gravitational energy, a spacequake surging through the cosmos at the speed of light. A fraction of a second before the detonation, the core of the star had been suddenly compressed into a compact ball some 10 miles wide, an incredibly dense mass in which a thimbleful of matter weighs up to 500 million tons (roughly the combined weight of all of humanity). This is the birth of what astronomers call a neutron star. Jolted by such a colossal stellar collapse, space itself was likely shaken—and shaken hard. The resulting ripples would have rushed from the dying star as if a giant cosmic pebble had been dropped into a space-time pond. These gravitational waves, though growing ever weaker as they spread from the stellar explosion, would have squeezed and stretched the very fabric of space-time itself. Upon reaching Earth they would have passed right through, compressing and expanding, ever so minutely, the planet and all the mountains, buildings, and people in their wake.

As shown by the example of the Sanduleak star, gravity waves are generated whenever space is fiercely disturbed. These waves are not really traveling *through* space, say in the manner a light wave propagates; rather, they are an agitation *of* space itself, an effect that can serve as a powerful probe. Light beams are continually absorbed by cosmic debris—stars, gas clouds, and microscopic dust particles—as they roam the universe. Gravity waves, on the other hand, travel right through such obstacles freely, since they interact with matter so weakly. Thus, the gravity wave sky is expected to be vastly different than the one currently viewed by astronomers. More than offering an additional window on space, gravity waves will provide a radically new perception. In addition, they will at last offer proof of Einstein's momentous mental achievement—their existence will demonstrate, in a firm and vivid way, that space-time is indeed a physical entity in its own right.

A few pioneering investigators in both the United States and Europe claimed to have detected the Sanduleak star's faint space-time rumble; other scientists are certain these claims are wrong. But the next time a gravitational wave rolls by, researchers are determined not to be caught off guard. Hence, the construction of LIGO. LIGO itself consists of two observatories, one in Louisiana and a near-twin in Washington state, which operate in concert with each other. But they are not alone. Similar instruments of varying size are being built in Italy, Germany, Australia, and Japan as well. These groups around the world are turning on the most sophisticated gravitational wave detectors to date, in hope of snaring their elusive cosmic prize. Almost no one doubts that gravitational waves exist, for there is already powerful evidence that such waves are real. Two tiny neutron stars—supernova remnants—in our galaxy have been observed rapidly orbiting each other. They are drawing closer and closer together. The rate of their orbital decay—about 1 yard per year—is just the change expected if this binary pair is losing energy in the form of gravitational waves. Direct reception of a wave, though, would offer the ultimate proof and provide astronomers with the most radical new tool in four centuries with which to explore the heavens.

In the early 1600s Galileo Galilei, then a savvy professor of mathematics at the University of Padua in Italy, pointed a newfangled instrument called a telescope at the nighttime sky and revealed a universe with more richness and complexity than previous observers ever dared to contemplate. The ancients had said that the heavens were perfect and unvarying, but Galileo discovered spots on the Sun and jagged mountains and craters on the Moon. With bigger telescopes came bigger revelations. We came to see that the Milky Way was not alone but one of many other "island universes" inhabiting space. More than that, all of these galaxies were rushing outward, caught up in the expansion of space-time. And as astronomers were able to extend their eyesight beyond the visible light spectrum and detect additional electromagnetic "colors," such as radio waves, infrared waves, and x rays, the heavens underwent a complete renovation. Long pictured as a rather tranquil abode, filled with well-behaved stars and elegant spiraling galaxies, the cosmos was transformed into a realm of extraordinary vigor and violence. Arrays of radio telescopes, aimed toward the edge of the visible universe, observed young luminous galaxies called quasars spewing the energy of a trillion suns out of a space no larger than our solar system. Focusing closer in, within our own stellar neighborhood, these same radio telescopes watched how neutron stars—city-sized balls of pure nuclear matter, the collapsed remnants of massive stars—spin dozens of times each second. Meanwhile, x-ray telescopes discovered huge amounts of x-ray-emitting gas, unobservable with optical telescopes, hovering around large clusters of galaxies. The invisible became visible.

The twenty-first century will certainly see the sky remade once again. It will happen as soon as astronomers detect gravity waves. These ripples in space-time will be seen neither with the eye nor as an image on an electronic display, not in the same way that visible light waves, radio waves, or x rays are distinguished. Each gravity wave that passes by Earth will, in a way, be felt—perceived, perhaps, as a delicate vibration, a vibrant boom, or even a low-key cosmic rumble. And when that happens, astronomy will never be the same again. It's as if in studying the sky we've been watching a silent movie—pictures only. Since their frequency happens to fall into the audio range, gravity waves will at last be adding sound to our cosmic senses, turning the silent universe into a "talkie," one in which we might "hear" the thunder of colliding black holes or the whoosh of a collapsing star. Firm discovery of these waves will at last complete the final movement of Einstein's unfinished symphony.

A gravitational wave telescope essentially acts like a geological seismometer but a seismometer that is placed on the fabric of space-time to register its temblors. The oldest detectors are designed as car-sized, cylindrical metal bars, capable of "ringing" like bells whenever a sizable gravity wave passes through them. The newest instruments, such as LIGO, involve a set of suspended weights that will appear to sway as the peaks and troughs of the traversing gravitational wave alternately squeeze and stretch the space between the masses (although these movements will be exquisitely tiny, the displacement being thousands of times smaller than the width of an atomic nucleus). Together, these various observatories will act as surveyor's stakes in pinpointing the source on the sky; by carefully monitoring the differing times that a gravity wave arrives at the detectors set around the globe, astronomers will be able to determine where the wave originated. The gravity wave signal could be regular or erratic, unceasing or sporadic. We would discern, in essence, a cosmic symphony of beats. And gravitational wave astronomers will translate these syncopated rhythms—the whines, the bursts, the random roars—into a new map of the heavens, a clandestine cosmos currently impossible to see.

This entire endeavor began very modestly in the 1960s, as one man's quixotic quest. At the University of Maryland, physicist Joseph Weber cleverly devised the first scheme to trap a gravitational wave and reported a detection in 1969. Inspired by Weber's insight, others quickly joined the campaign. Gravitational wave detectors were erected around the world. In the end, Weber's sighting was never confirmed beyond dispute. Indeed, many contend his evidence has been refuted. But that didn't deter the newcomers to gravitational wave physics from continuing the search. They were energized by the technological challenges of the problem. Weber triggered a movement whose momentum has never diminished to this day. A host of specialists—in optics, lasers, materials science, general relativity, and vacuum technology—have now come together to produce the most complex instruments ever devised for an astronomical investigation. With no guarantee that a signal will be detected, critics have fiercely argued that this attempt is being made too soon. Many in the astronomy and physics communities waged a strong campaign against the endeavor, declaring that the money would be better spent on surer scientific quests. But the potential of the science—not to mention strong politicking—overrode those concerns. As a result, gravitational wave researchers are not just carrying out an experiment, they are founding a field. The questions they are asking stretch back to Aristotle, and answers may at last be within their reach. *Einstein's Unfinished Symphony* will show how these efforts are the culmination of a centuries-long pursuit—nothing less than the ambition to unravel the enigma of space and time.

Space in G-Flat

*W*e talk about space so readily and easily: "There's no space for an office in this apartment" or "Give me some space, man." The concept of space seems self-apparent to the casual observer. But on deeper reflection its true nature remains elusive. "As a rule, people differentiate between matter, space, and time. Matter is what exists in space and endures through time. But this does not tell us what space is. . . . It is matter which we see, touch, and hear, which causes sensations to arise within us. . . ," the British philosopher Ian Hinckfuss once noted. Space is perceived yet not felt, observed, or heard. So what, after all, is space?

Perceiving the limits and bounds of space was probably one of *Homo sapiens*' earliest accomplishments. Space was first and foremost an orientation among familiar objects, discerning the effort needed to reach a nearby river, rock, or tree. A sense of space may even have preceded a perception of time, since we describe moments of time as "short" or "long," words that usually describe spatial categories. And with the emergence of agriculture came the need to measure space exactly for practical considerations, such as planting a field or digging an irrigation ditch.

From these humble beginnings arose a more esoteric contemplation of space. In ancient Greece philosophers developed the concept of the void—the *pneuma apeiron*—a vacancy that allowed for a separation of things. Democritus, father of the atomic theory, required such a void—a nothingness bereft of matter—for his idea to work. Space was the empty extension that allowed his bits of matter—his atoms—to move about. Such discussions soon extended to thinking about space in the abstract. A Greek philosopher named Archytas asked what would happen if you journeyed to the end of the world and stretched out your hand. Would your hand be stopped by the boundaries of space? Lucretius, Democritus's pupil, answered *no*, and he had an interesting proof for space being unbounded. He asked: Suppose a man runs forward to the very edge of the world's borders and throws a winged javelin. Unable to conceive that anything could get in the javelin's way, Lucretius concluded that the universe must stretch on and on without end. Aristotle, on the other hand, took the opposite stand. He stated that it was "clear that there is neither place nor void nor time beyond the heaven." For him the universe was finite. If a stone falls to the Earth to find its natural place at the center of the universe, he argued, then fire moving upward in the opposite direction must also face a limit. In Aristotle's physics, upward and downward motions had to be balanced. Moreover, any objects in the farthest reaches of an infinite universe, forced to rotate about a motionless Earth, would end up traveling at infinite velocities, a situation that Aristotle considered patently absurd.

Space was the subject of fierce intellectual debates into the Renaissance. In medieval times theological concerns often prejudiced the debate. To think of an immovable void, as outlined by the Greek atomists, meant God created something He could not budge. Such a situation challenged His omnipotence. Thus, this idea was deemed heretical and was avoided. But as natural philosophers, starting as early as the fourteenth century, began to consider kinematics, the motion of objects, it became necessary to contemplate some kind of fixed space. They needed to imagine a special motionless container in order to understand such physical concepts as velocity and acceleration. One man in particular would change the landscape of science in making this assumption as he searched for the mathematical rules by which motion could be predicted. That man was Sir Isaac Newton.

The dreaded Black Death appeared in London in 1665. With the plague spreading northward to the university town of Cambridge, Newton fled that summer to his childhood manor home, Woolsthorpe, in Lincolnshire. In that rural setting he worked intensely for two years. Still in his early twenties, he was laying down the mathematical and physical foundations of his most important ideas, which had been germinating throughout his college years: the theory of color, the construction of the calculus, and, most importantly, the laws of gravitation. He returned to Cambridge in 1667, at the age of 24, and within two years became Lucasian Professor of Mathematics, a high honor at the university. Secretive, obsessive, and fearful of exposing his work to criticism—a man chockful of neuroses—Newton let many of his revolutionary thoughts go unpublished. It was not until 1684, sparked by the questions and persistent prodding of Edmond Halley (of comet fame), that Newton was at last convinced to write his masterpiece, *Philosophiae naturalis principia mathematica* (Mathematical Principles of Natural Philosophy). He abandoned his work on alchemy, his most recent fascination, and applied his legendary power of concentration completely on the *Principia* (as it is most familiarly known) for nearly two years.

The *Principia* deals with both gravity and the mechanics of motion. Forces in nature, declared Newton, are not needed to *keep* things moving (as Aristotle argued); rather, forces *change* motion and in predictable ways. Newton clarified what Galileo had begun to infer from experimental tests: an object in motion does not naturally come to a stop; instead, it will remain in motion unless altered by an outside force, such as friction. The effect of a force is to get an object moving, to stop it, or to change its direction. And when it comes to gravity,

Newton revealed that the strength of the gravitational attraction between two objects depends on two things: the total amount of matter in each object and the distance between them. The greater the mass of each object, the stronger the pull; conversely, the larger the separation between the two masses, the weaker the attraction. Or as Newton put it, two objects exert a gravitational force on one another that is in direct proportion to their masses and in inverse proportion to the square of their distance. More importantly, he realized that what draws an apple to the ground (an event he presumably witnessed as a young man at Woolsthorpe during his early ruminations on gravity) also keeps the Moon in orbit about the Earth. Moreover, he deduced the exact equations for determining those motions. Newton had discovered that nature uses a mathematical treatise as its playbook. As if with one monumental stroke of the pen, he established that motions everywhere, in the heavens and on Earth, are described by the same physical laws. Before this insight, philosophers generally believed that the heavens were distinctly different from the domain of man. Earthly things were mortal—subject to change and transition—while the stars and planets were eternal and incorruptible. But with Newton's new laws the cosmos and terra firma were blissfully wedded. An all-encompassing set of mathematical rules could now explain events in both domains: the ocean tides, the motion of comets and planets, as well as the projectile paths of cannon balls. All these phenomena could be tracked with the same clocklike precision. So great was this achievement that Newton was the first person in England to be knighted for his scientific work.

Skydivers and bungee jumpers, plummeting toward the ground, have great respect for the force of gravity pulling them downward. Gravity is also the ruling force in determining the universe's evolution and its grand structure. Yet gravity is the weakest force in the cosmos. It seems paradoxical. A toy magnet can easily pick up a paper clip against the gravitational pull of the entire Earth. Or take two protons sitting next to each other. The gravitational force between them is a trillion trillion trillion times weaker than the electrostatic force pushing on them. Gravity gains collective strength only when masses accumulate and exert their effect over larger and larger distances. In this way gravity comes to control the motions of planets, stars, and galaxies.

Space and time loom large in Newton's laws, for laws need a framework. Take Newton's first law of motion. A body either remains at rest or in continuous uniform motion (traveling in a straight line at a constant speed), unless an outside force causes that state to change. But at rest in relation to what? Or in motion to or away from what location? As soon as one talks of "motion," one must establish a home base. Think of a child reading a book in a moving car. To someone on the curb watching the car whiz by, the book is moving fast. To the child inside, it is perfectly still. Newton's critical choice was to establish a reference frame in the universe at large. Space itself became his motionless laboratory: flat, penetrable, yet forever the same. He was not the first to think this way—Galileo, for one, posited a continuous three-dimensional void—but Newton made it an integral component of the Western canon. "Absolute space in its own nature, without relation to anything external, remains always similar and immovable," he authoritatively stated. Space was at rest, and everything else in the universe moved with respect to that. To Newton, space was an empty vessel. You were either at rest or in motion with regard to this container. Positions, distances, and velocities were all measured in regard to this fixed space. Only by establishing this framework—this unchanging cosmic landscape—could his equations work.

Measuring motions in this absolute space also required a universal clock, which ticked off the seconds for all the inhabitants of the cosmos. Events everywhere, from one end of the universe to the other, were in step with the ticks of this grand cosmic timepiece, no matter what their speed or position. A clock sitting at the edge of the universe or zipping about the cosmos at high speed would register the same passage of time, identical minutes and identical seconds, as an earthbound clock. This meant that two cosmic observers, perched on opposite sides of the universe, could synchronize their watches instantaneously. Moreover, "the flowing of absolute time is not liable to any chance," said Newton in the *Principia*. His clock was never affected by the events going on around it. Like some cosmic Big Ben, time stood aloof, as galaxies collided, solar systems formed, and moons orbited planets in this vast universe of ours.

Newton's law of gravity, brilliant in its ability to predict the future paths of planets, did have an Achilles' heel. It provided no explanation for the mechanism underlying gravity. There was no medium or physical means to push and pull the planets and other objects around. Newton's tendrils of gravitational force just appeared to act instantaneously over vast distances, as if by magic. This feat appeared more resonant with the occult. As one wry critic of the time noted, "Newton calculated everything and explained nothing." For some, the lack of a cause was tantamount to bad science. Newton was aware of these difficulties and lamented that one body acting on another through a vacuum, "without the mediation of anything else, by and through which their action and force may be conveyed from one to another, is to me so great an absurdity, that I believe no man who has in philosophical matters a competent faculty of thinking, can ever fall into it." But his decision to stand by his laws was a practical one. He chose the path that allowed successful predictions to be made. As Einstein said in an imaginary talk with Newton, "You found the only way which, in your age, was just about possible for a man of highest thought and creative power." Newton's decision to define an absolute space and absolute time was flawed, but it became ingrained into the very fabric of physics simply because it came up with the right answers.

Newton's conception of an absolute space and absolute time would influence the course of physics for some 200 years, but it was not universally accepted. There were critics who raised their voices loudly. The most notable were the British philosopher George Berkeley and the German diplomat and mathematician Gottfried Wilhelm Leibniz, who was Newton's archrival for his claim to have invented the calculus first. To them space and time were not fixed entities at all. Leibniz declared that "space and time are orders of things, and not things." Space and time could only be defined with regard to their relation with objects of matter. For Muslim philosophers, who had developed similar theories, this avoided the question, "Where was God before the creation?" The answer was simply that there was no "place." Space did not exist until the creation of matter. Toward the end of his life, Newton found solace for his worries about absolute space and time in a religious explanation: "[God] endures forever, and is everywhere present; and by existing always and everywhere, he constitutes duration and space." Newton's perception would indeed endure and for one simple reason: his equations worked. Mathematics had largely been an aesthetic experience for the Greek philosophers. Newton changed that attitude by demonstrating, with his powerful law of gravitation, that mathematics offered a road to discovery. He transformed mathematical law into physical law, rules by which the operations of nature—planetary motions, the propagation of light, mechanics—could be predicted. Eventually belief in the infallibility of these laws grew so strong that when Newton's equations appeared to fail (as in the case of explaining the orbital motions of the planet Uranus), it was immediately assumed that the equations were still valid and that an unseen planet was lurking beyond Uranus to account for the discrepancy. In that particular case, such resolute faith paid off handsomely. Adherence to Newton led to the discovery of the planet Neptune in 1846. It was difficult for Newton's critics to fight such success.

In his mathematical choices Newton assumed that space was "Euclidean," holding all the properties defined by the famous Greek geometer Euclid in the third century B.C. Although the rudiments of geometry had been fashioned in Egypt along the banks of the Nile River—knowledge gained by the Pharaoh's surveyors, the *harpedonaptai* or rope-stretchers—those rules turned into a mathematical discipline when they traveled to Greece. The Greek philosophers saw in geometry a pure set of truths, which could be arrived at through logic alone. Geometry was proof that knowledge of the physical world could be gleaned through pure reason. So revered was geometry that when Plato established his Academy it was said that the sign over the door announced, "Let no one enter here unless he knows geometry." Euclid stood at the zenith of this movement. Around 300 B.C. he wrote *The Elements*, which expressed all geometrical knowledge then known as a concise set of axioms and postulates. It served as the basis for all mathematical thought for the next 2,000 years.

In this major work Euclid defined a space that was flat, which is exactly the world we perceive and measure around us when confined to the ground. He also made a list of all the geometric ideas we take for granted—for example, that a straight line can be drawn between any two points or that all right angles are equal to one another. These were self-evident truths. His fifth postulate considers a line and any point not on that line. According to the ancient Greek geometer, there is one—*and only one*—line that can be drawn through that point that is parallel to the original line. The two lines, like two parallel railroad tracks, will never meet. To our senses there appears to be no other possible configuration. While it seems self-apparent that parallel lines will never converge, later mathematicians were caught up in studying this particular axiom in more depth. Rather than assuming it was a given, they wondered whether the rule could be derived directly from Euclid's other axioms. They wanted to prove it explicitly, rather than just state it as true.

One tried-and-true mathematical trick to test a postulate is to assume it is false and see what happens. That's precisely what a Jesuit priest named Girolamo Saccheri did in 1733. He assumed that the parallel axiom was false and then showed it would lead only to absurdities—hence, the name of this technique, *reductio ad absurdum*. Saccheri discovered that he could get *more* than one line through a point to be parallel to a given line. Figuring this was clearly ridiculous, he concluded that he had proven what he had set out to do—show that the axiom was clearly true as stated so elegantly by Euclid. Saccheri failed to see that he had accidentally stumbled onto a whole new geometry.

By 1816, after years of thinking about the problem, another mathematician arrived at the same insight and backed off as well but this time out of fear of ridicule. The great German polymath Carl Friedrich Gauss uncovered the same absurdities that Saccheri did. He did not reject them outright, though; he just knew that such challenges to the great Euclid would be considered heresy. As a result, Gauss never officially reported his findings during his lifetime (although he did discuss the new geometry he developed in private correspondence with colleagues). A reclusive man unwilling to start a public dispute that would disrupt his peace of mind, Gauss carefully guarded his secret, fearing as he put it, "the clamor and cry of the blockheads" over questioning mathematics' sacred gospel. Euclid's framework, sturdy for centuries, served as the very foundation of mathematics. Gauss was also a perfectionist, who kept much of his work to himself. He was terribly reluctant to publish any idea until he had polished its proof to a fine sheen. It's not surprising that his personal seal, a tree with sparse fruit, bore the motto *Pauca sed matura* ("few but ripe").

Realizing that non-Euclidean geometries were a possibility (at least on paper), Gauss gradually began to wonder whether a non-Euclidean geometry might describe true physical space. Perhaps space wasn't flat, as Newton assumed, but rather curved. His musings were amplified by a practical concern. He was commissioned by his government in the 1820s to conduct a geodetic survey of the region around the city of Göttingen in Hannover. This endeavor enhanced the thoughts he was already having about curved space. Curvature, he realized, need not be restricted to two dimensions, such as the rounded surface of a planet. In an 1824 letter to Ferdinand Karl Schweikart, a professor of law and a geometer as well, Gauss bravely mentioned that space itself, in all its three dimensions, might be curved or "anti-Euclidean" as he called it. "Indeed," he wrote, "I have . . . from time to time in jest expressed the desire that Euclidean geometry would not be correct." He may even have tested this hypothesis during his geodetic work. Using light rays shining from peak to peak in the Harz Mountains, Gauss had surveyed a triangle of pure space formed by three mountains, the Hohenhagen, the Brocken, and the Inselberg. By his metric figuring, the sides of the triangle measured 69, 85, and 107 kilometers. No deviation from flatness, however, was detected.

Others were more open in exploring this new geometric terrain. Between 1829 and 1832, while Gauss kept mum publicly from his faculty post at the University of Göttingen, two mathematicians independently published papers stating that it was possible to have geometries that disobeyed Euclid. One proof was done by the Russian mathematician Nikolai Lobachevsky, the other by the Austro-

Hungarian János Bolyai, allegedly the best swordsman and dancer in the Austrian Imperial Army in his day. Like Saccheri a century earlier, Lobachevsky and Bolyai had asked what would happen if the fifth postulate were wrong. If it were in error, what type of mathematics arises? What if it is assumed that an infinite number of lines can be drawn through a point near a given line without any of the lines intersecting? In this way the two mathematicians came to describe a space of *negative* curvature.

"From nothing I have created another entirely new world," wrote Bolyai to his father, who had struggled with the fifth postulate himself when he was a student friend of Gauss. Bolyai's new world can be visualized by imagining a triangle drawn on the surface of a saddle. The triangle, with its sides curved inward, would appear a bit shrunken. So the sum of its angles is not 180 degrees, as authoritatively stated in high school textbooks as the standard Euclidean answer. Instead, it is less than that. Being concave, the saddle's surface also allows many lines, which never meet, to be drawn through a point near a given line. Lobachevsky called this new system his "imaginary geometry."

Like Gauss, Lobachevsky also thought of the possibility that three-dimensional space might be curved but figured that distances far longer than the spans between alpine mountains would be needed to test such a radical idea. He suggested conducting certain parallax measurements on distant stars. When the measurements were carried out, no change from flatness was discovered. Hence, it was assumed that Euclid's rules continued to reign supreme throughout the universe.

Meanwhile, Gauss's fascination with the new geometry was passed on to a brilliant student at the University of Göttingen, Bernhard Riemann, who developed another non-Euclidean geometry altogether. The new construct was revealed during a trial lecture the timid 27-year-old student gave while seeking appointment as a *Privatdozent* (lecturer) at the university in 1854. In the course of his lecture, prepared in only seven weeks and later described as a high point in the history of mathematics, Riemann introduced a geometry in which *no* parallel lines can be drawn through a point near a given line. He was dealing with a space of *positive* curvature, best typified by the surface

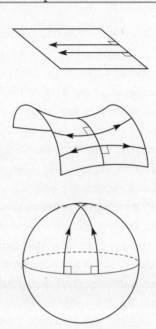

Three different geometries: flat space (top); negatively curved space (middle); and positively curved space (bottom).

of a sphere. Here, the shortest distance between two points is not a straight line; instead, the shortest path is an arc, a segment of a great circle that encompasses the entire sphere. Like the lines of longitude on Earth, each great circle eventually intersects with every other great circle at the poles of the sphere. Consider two adjacent lines at the equator aligned directly north to south. Locally, they seem as parallel as they can be. Yet extend these lines around the world and they eventually cross and meet. Consequently, there are no parallel lines in this special kind of geometry. A triangle on such a curved surface would look inflated; the sum of its angles would be more than 180 degrees. Like Gauss, Lobachevsky, and Bolyai before him, Riemann was discovering that a mathematician can imagine many different geometric worlds. Euclid did not corner the market after all.

Infamous for his stern and critical demeanor, Gauss displayed a rare enthusiasm at the end of Riemann's presentation. He was perhaps the only one in the audience that day who recognized that Riemann had surpassed his predecessors by extending non-Euclidean geometry much farther. Riemann generalized the geometry of curved spaces to higher dimensions, spaces involving four, five, and even more dimensions. While at the time these manipulations may have appeared to be no more than a mathematical game, this work would later prove invaluable when Einstein faced the awesome task of developing his general theory of relativity. Riemann was fashioning the tools that allowed Einstein to envision a completely different view of space and time. Riemann served as a vanguard for the Einsteinian revolution to come. At one point he dared to suggest that the true nature of space would not be found in ancient manuscripts from Greece but rather from physical experience. He even imagined that the universe might close in on itself, forming a sort of four-dimensional ball. Such a curvature would be noticed only over great distances, hence our common experience of seeing our local universe as flat. Interestingly, Riemann went on to consider whether the structure of space was somehow molded by the presence of matter, creating what he called a *metrical* field akin to an electromagnetic field. It was a prescient vision, but he spoke too soon. Physics was not yet ready to give up on its pleasant Newtonian world, a world of absolute and rigid space, unvarying and unchanging. The idea that space might display a distinct geometry elicited fury in certain philosophers of that era. Space was still considered an empty vessel devoid of physical properties.

Riemann's life was tragically cut short. He died of tuberculosis at the age of 39 in the village of Selasca on Lake Maggiore. He had gone to Italy to attempt a cure. One of Riemann's greatest desires was to unify the laws of electricity, magnetism, light, and gravity. Such a project was premature, but his mathematics would still become the vital ingredient of the new physics to come. "Riemann left the real development of his ideas in the hands of some subsequent scientist whose genius as a physicist could rise to equal flights with his own as a mathematician," said mathematician Hermann Weyl. After a lapse of 49 years, that mission would at last be fulfilled by Einstein. Had Riemann lived to a ripe old age, it is conceivable that Einstein would have thanked him in person.

The Maestro Enters

The tale has been told and retold so many times that it has taken on the strains of a fable. In some autobiographical notes Einstein remembered being haunted as a lad by a strange musing: If a man could keep pace with a beam of light, what would he see? Would he observe a wave of electromagnetic energy frozen in place like some glacial swell? "It does not seem that something like that can exist," he recalled thinking at the youthful age of 16. Here was the seed for Einstein's casting out the absolute space and absolute time of Isaac Newton. Relativity would arrive, not from concerns over the flaws in Newton's mechanics, but rather from contemplating the forces of electricity and magnetism as well as the mysteries of light.

For most of history it was generally assumed that light was something that was transmitted instantaneously. In a way it was everywhere "there." With this premise the light from a far-off star arrived at our eyes on Earth as soon as it was emitted. By the seventeenth century, however, certain thinkers began to wonder whether light had a finite speed after all—like sound—only far, far faster. In his great treatise on mechanics in 1638, Galileo may have been the first to suggest an experiment to test this hypothesis directly. One man stands on a hill and uncovers a lantern, signaling a companion positioned on another hill less than a mile away. The second man, as soon as he sees the initial beacon, flashes a return light of his own. Performing this task on hills spaced farther and farther apart, Galileo figured the men would spot a successively longer delay between the dual flashes, which would reveal the speed of light. This test was eventually carried out by members of the Florentine Academy. Of course, no delay was detected, given the crudeness of Galileo's experiment. Human reaction time is far too slow. The vast span of our solar system provided a far better test.

In the 1670s—Newton's day—the Danish mathematician and astronomer Ole Römer closely studied the movements of Jupiter's four largest moons, particularly the innermost one, Io. Specifically, he carefully monitored the moment when Io periodically moves behind Jupiter and gets eclipsed. In doing this, he noticed that the interval between successive eclipses (an event that occurred about every 42 hours) was not constant but regularly changed, depending on the position of Earth in relation to Jupiter. When Earth was moving away from Jupiter in its orbital motion around the Sun, the expected moment for Io to eclipse arrived later and later. This is because the light bringing that information to your eyes has to travel a bit more distance with each eclipse. By the time Earth reached its farthest point from Jupiter, Römer's measured delay mounted up to 22 minutes (a better figure is 16.5 minutes). Others had noticed such changes before, but Römer shrewdly deduced that the delay was just the time needed for Io's light to traverse the extra width of Earth's orbit. Dividing Earth's orbital width (186 million miles) by the delay time, Römer's crude measurements pegged a light speed of around 140,000 miles per second. That's quite fast, but Römer had shown it was certainly not instantaneous. The modern value is 186,282 miles per second.

By the nineteenth century physicists made great strides in understanding the nature of light. It was the era when scientists verified that

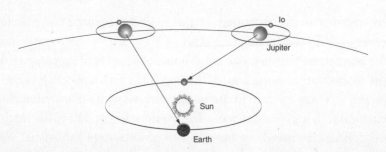

When Earth is farthest from Jupiter, the eclipse of Io is sighted later than expected because the light must travel the extra width of Earth's orbit. In the seventeenth century, Ole Römer used this effect to make the first good estimate of the speed of light.

light behaved like a wave. And, according to the physics of the 1800s, that required the light wave to be propagating through some kind of medium. No reputable physicist would have dared to imagine that two objects could transmit light between one another lacking a substance to carry the wave. Sound waves move through air, and ocean waves travel through water; if there is a wave, *something* must be waving. For the heavens the transmitting agent came to be known as the "luminiferous ether," a reworking of the heavenly ether once postulated by the ancient Greeks. It filled the whole universe. As conceived, the ether was a rather odd material: it had to be rigid enough to transmit light waves at terrifically high speed yet still allow the Earth, planets, and stars to move through it without resistance. Such a paradox provided theoretical physicists with a cottage industry for more than a century. Scientific journals were filled with attempts to explain how the ether could be both stiff and insubstantial. Permeating all of space, the ether also served as a motionless reference system. In a way the ether resembled a vast body of water. As a wave passes through the open sea, the water moves only up and down. The wave is transmitting energy but not moving the water forward. The same was thought to be true for the ether. Here at last was Newton's absolute rest frame in physical form.

At the same time, explorations were taking place into the nature of electricity and magnetism. A link between these two phenomena first came in 1820 when the Danish physicist Hans Christian Ørsted discovered that an electric current in a wire deflects a compass needle; in other words, a conducting wire acts as a magnet. Stop the current and the magnetism vanishes. The Englishman Michael Faraday completed the connection when he noticed the opposite effect: a moving magnet creates an electrical current. Born into poverty, with no formal mathematical training, Faraday was a self-taught scientist. His mathematical deficiencies may have been to his advantage. Highly visual, he imagined his magnetic objects being surrounded by fields of force, invisible lines influencing the movements of objects within their midst. Such fields are wonderfully displayed in the way iron filings align themselves when you sprinkle them around a magnet. Likewise, Faraday thought that electric fields somehow manipulated charged particles with ghostly hands.

The distinguished Scotsman James Clerk Maxwell was just one in the legion of people, both scientists and nonprofessionals alike, who were captivated by Faraday's experiments. A handsome man with a delicate constitution, Maxwell became a professor of natural philosophy at the age of 24. Ten years later, with his august *Treatise on Electricity and Magnetism*, a triumph of nineteenth-century physics, he composed the mathematical "words" to explain Faraday's fields of force. His derivations, a set of four partial differential equations so elegantly succinct that they show up on physics students' T-shirts, demonstrate how electricity and magnetism, two forces that on the surface seem so disparate, are merely two sides of the same coin, each unable to exist in isolation. Maxwell united them into the single force of electromagnetism.

More than that, Maxwell's equations also revealed that an oscillating electric current—charges moving rapidly back and forth—would generate waves of electromagnetic energy coursing through space. He even worked out the speed these waves would travel, which was related to the ratio of certain electrical and magnetic properties. The result turned out to be exactly equal to the speed of light. Was this just a coincidence, as some thought? Maxwell boldly said *no*. He concluded that light itself was a propagating wave of electric and magnetic energy, an undulation that moved outward in all directions from its source.

Visible light waves, each measuring about 1/50,000 of an inch from peak to peak, were just a small selection of the wide range of waves possible. There could also be waves of electromagnetic energy both shorter and longer than visible light. The German physicist Heinrich Hertz proved this very fact in 1888. In a laboratory humming with spark generators and oscillators, Hertz created the first radio waves. Each wave had a length of 30 inches and sped across his lab at the speed of light. It was the first experimental verification of Maxwell's prediction that electromagnetic waves existed.

Maxwell died of abdominal cancer in 1879 at the age of 48, nine years before Hertz conducted his experiments. In the year before his death Maxwell also wondered about the problem that plagued physicists through most of that century: the motion of the Earth through the ether. Given the assumption that Earth was moving through a stationary ether, Maxwell thought of an optical experiment to detect an "ether wind" as Earth sailed at some 67,000 miles per hour in its annual voyage around the Sun. Like the air rushing past the passengers in a speeding convertible with its top down, the ether would be blowing past the Earth. Sparked by Maxwell's challenge, a young U.S. naval officer, Albert A. Michelson, built a special instrument to spot the ethereal breeze in 1881 while assigned to Berlin for postgraduate work in physics. The instrument was his own design, an intricate assembly of mirrors and prisms that allowed pencil-thin beams of light to bounce back and forth in order to peg light's speed. It came to be known as the Michelson interferometer. Michelson found no hint of a draft, though. Berlin traffic outside his laboratory rattled the detector at times, hurting its sensitivity. He tried again in 1887 as a civilian professor at the Case School of Applied Science in Cleveland, Ohio. There he had teamed up with Edward Morley, a chemist at neighboring Western Reserve University. Using a vastly improved interferometer, the two researchers sent one beam of light "into the wind" in the direction of Earth's orbital motion and directed another beam at right angles to this path. Michelson once explained to his young daughter Dorothy how the test worked: "Two beams of light race against each other, like two swimmers, one struggling upstream and back, while the other, covering the same distance, just crosses and returns." The light beam fighting the ethereal "current" was expected to move a bit slower.

Michelson and Morley's elaborate equipment, set up in a base-ment laboratory, was mounted on a massive sandstone slab that floated on a pool of mercury to cut down on vibrations. But even with such precautions the two scientists detected no difference whatsoever in the measured velocity of their two beams of light, no matter which way the beams were pointed. The technique was so sensitive that it should have been able to measure a wind speed as small as 1 or 2 miles a second. With the Earth moving more than 10 times faster, Michelson had figured they would easily spot it. He was dismayed to discover that he was wrong. In 1907 Michelson became the first American to win a Nobel Prize in the sciences. He was honored for the development of his exquisitely sensitive optical instruments, many of them inspired by his futile search for the ether.

Others looked for the ether and failed as well. (Even Einstein, as a college student, wanted to build his own apparatus to measure the Earth's movement against the ether, but a skeptical teacher nixed the plan.) The null results forced physicists to come up with elaborate schemes to explain the lack of a detectable "wind." First the Irish physicist George FitzGerald and later Hendrik Lorentz, a Dutch physi-cist, suggested that an object traveling through the ether would con-tract—get physically compressed—in the direction of the motion. The dimensions of an object would somehow change as they moved about. That could explain why Michelson didn't notice a change; it was neatly canceled by this effect. Eventually, the great French mathematician Henri Poincaré railed against these complicated efforts to explain why physicists were not detecting Earth's absolute motion through the ether. In 1904 he presciently spoke of the need for a "principle of rela-tivity." Soon such a theory would be provided.

Historians continue to debate whether the Michelson-Morley experiment influenced Einstein in any way. He makes only a brief and indirect reference to the test in his famous 1905 paper on relativity, even though his new theory neatly explained Michelson's continuing failure to discern the ether. What Einstein did stress was his perplexity over certain properties of electricity and magnetism. He was bothered by a seeming paradox. Consider either a bar magnet moving through a fixed coil or a coil moving over a stationary bar magnet. Each case is distinct. Maxwell's equations must handle each situation differently, depending on whether the coil is stationary and the magnet is moving or the magnet is stationary and the coil is moving. But each case leads to the exact same result: a current. Why should that be, asked Einstein? The descriptions of what is happening are different for the two different points of view, yet the observed outcome—the flow of an electric current in the coil—is the same. The experiment cannot reveal which object—the coil or magnet—is *really* moving in absolute space. Here was a crack in the notion of a fixed, eternal reference frame.

Both Newton's mechanics and Maxwell's equations of electromagnetism were the two monumental theories of their era. Each yielded extremely accurate predictions. What disturbed Einstein was that these two great works of physics didn't seem to share the same rules on defining a space and time. Einstein's masterstroke was finding a way to make the two theories compatible with the simplest assumptions possible. Perhaps surprising was that his solution required no grand leaps of physics. Einstein's historic 1905 paper is exquisite in its simplicity. All of his hypotheses are based on the physics available in the nineteenth century. His one inventive assumption was a new conception of space and time. With that one change, all fell into place.

The prevailing image of Einstein has long been that of the venerable elder, the Chaplinesque figure with baggy sweater and frightwig coiffure. But the youthful Einstein, at the height of his scientific prowess during the development of relativity, was a man whose limpid brown eyes, wavy hair, sensuous mouth, and virtuosity on the violin aroused considerable attention, especially among women. One acquaintance compared Einstein's demeanor to that of a young Beethoven, full of life and laughter. Yet like the great romantic composer, the twentieth-century's most celebrated scientist had his dark side as well. He was also a loner (despite two marriages), a sharptongued cynic at times, and a self-centered man who could serve humanity yet express little empathy for the problems of those close to him. Einstein was born in 1879, the year of Maxwell's death, into a nonreligious Jewish family, one well assimilated into the culture of southern Germany for more than two centuries. His father ran, with mixed success, an electrical engineering company, the high-tech business of its time. Early on Einstein expressed an intense desire to learn things in his own way. He apparently didn't talk until the age of 3, stubbornly waiting until he could speak in complete sentences. Growing up with his younger sister Maja, little Albert loved doing puzzles, building structures, playing with magnets, and most especially solving geometry problems, the very key to his later work. He detested the German school system, with its emphasis on rote learning. And he didn't suffer fools gladly: he eventually dropped out of gymnasium (high school) due to conflicts with a teacher, among other reasons. Fortunately, he was able to enroll in a Swiss university, the Polytechnic in Zürich (the Federal Institute of Technology), although he was hardly a favored student among his professors. One called him a "lazy dog." As a result, he found no academic post upon graduation, surviving instead on the occasional teaching or tutoring assignment. Only in 1902 did he get a bona fide job, at the Swiss patent office in Bern. But all the while he was religiously devouring the books of the physics masters. He preferred self-learning. He had that spark—that devotion—since childhood.

Physics was at a critical juncture at the turn of the twentieth century. X rays, atoms, radioactivity, and electrons were just being discovered. It took a rebel—a cocky kid with mediocre college grades and no academic prospects but an unshakable faith in his own abilities—to blaze a trail through this new territory. Fearless at challenging the greats of his day, even as a student, Einstein was sure that the prevailing theory linking Newtonian mechanics with electromagnetism—electrodynamics—did not "correspond to reality . . . that it will be possible to present it in a simpler way." His work as a patent examiner turned out to be a blessing. He would fondly remember his government office as "that secular cloister where I hatched my most beautiful ideas." Unencumbered by academic duties or pressures, Einstein was able to explore those ideas freely. By 1905, at the age of 26, like a dormant plant that suddenly flowers, he burst forth with a series of papers published in the distinguished German journal *Annalen der Physik*. Any one could have garnered a Nobel Prize. Inspired by the new quantum mechanics, he first proposed that light consists of discrete particles, what came to be known as photons (the Nobel winner). Second, he explained the jittery dance of microscopic particles—Brownian motion—as the buffets of surrounding atoms. It helped persuade the scientific community that atoms truly exist. Lastly, he submitted a paper blandly entitled "On the Electrodynamics of Moving Bodies" in which he revealed his special theory of relativity (actually rejected as a doctoral thesis topic for being too speculative).

Many of Einstein's mathematical arguments were similar to those already used by Lorentz and Poincaré, but there was a vital difference. Unlike his predecessors, Einstein was redefining the nature of time itself. He remarked many years later that the subject had been his "life for over seven years." Special relativity proposed that *all* the laws of physics (for both mechanics and electromagnetic processes) are the same for two frames of reference: one at rest and one moving at a constant velocity. Einstein was saying that a ball thrown up into the air on a train moving at a steady 100 miles per hour on a straight track behaves just the same as a ball thrown upward from a motionless playground. For that to be true, though, means that the speed of light must also be the same in each environment, both on the train and on the ground. If the laws remain the same, each must measure the same speed of light. "[We will] introduce another postulate . . . ," wrote Einstein in his historic 1905 paper, "that light is always propagated in empty space with a definite velocity c, which is independent of the state of motion of the emitting body."

Let's make the comparison fairly drastic, since the effects of relativity are not really noticeable unless the comparative speeds are extremely high. Consider a spaceship racing by the Earth at a steady 185,000 miles per second, just under the velocity of light. Common sense might lead you to believe that the astronauts would be going nearly as fast as any light beam passing by, much as Einstein imagined as a youth. But that's not the case at all. The astronauts on that spaceship will still measure the velocity of a light beam at 186,282 miles per second, just as we do here on Earth. This situation seems bizarre, but not really. The speed of light remains constant, but other measurements get adjusted. The seeming paradoxes that arise are taken care of by acknowledging that time is not absolute. Time is, well, relative. The very term "velocity" (miles per hour or feet per second) involves keeping time, but the astronauts and Earthlings do not share the same time standard. That was Einstein's genius. He recognized that Newton's universal clock was a sham.

Since nothing can travel faster than the speed of light, two observers set apart in different frames of reference cannot really agree on what time it is. The finite speed of light prevents the two from synchronizing their watches. Einstein discovered that observers separated by distance and movement will not agree on when events in the universe are taking place. Consequently, by just looking, the Earthlings and astronauts will not agree on each other's measurements as well. Mass, length, and time are all adjustable, depending on one's individual frame of reference. Look from Earth at a clock on that spaceship that is whizzing past the planet. You will see time progressing more slowly than here on Earth. You will also see the spaceship foreshortened in the direction of its motion. Those on the spaceship, who perceive no changes in themselves or in their clock's progression, look at their home planet swiftly moving past them and see the same contraction and slowing of time in the Earthlings! Each of us measures a difference in the other to the same degree. Space shrinks and time slows down when two observers are uniformly speeding either toward or away from one another. Lorentz and FitzGerald spoke of an actual contraction in absolute space. Einstein, on the other hand, showed that the changes are a perception of measurement. Space and time will be different in each reference frame. The only thing that the Earthlings and astronauts will agree on is the speed of light in a vacuum.* It is the one universal constant.

With absolute time destroyed, there was also no need for absolute space either. Our intuition that the solar system sits serenely at rest, with the spaceship speeding away in some motionless container of space, no longer works. It could just as easily be the astronauts at rest, with the Earth speeding away. The "introduction of a 'luminiferous

---

*Light does appear to travel more slowly when transmitted through matter. Due to the light's interactions with the matter along the way, its overall speed is effectively reduced.

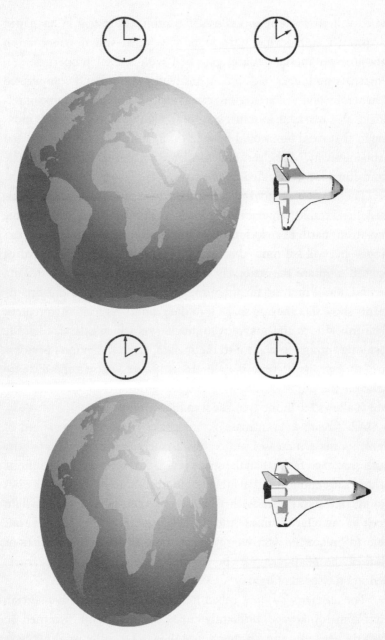

From Earth, a spaceship uniformly moving past the planet at near the speed of light will look shorter from end to end. Its clock will appear to tick more slowly. Those on the spaceship will perceive the same contraction and slowing of time on Earth.

ether' will prove to be superfluous," continued Einstein in his paper, "inasmuch as the view here to be developed will not require an 'absolutely stationary space' provided with special properties . . . ." Physicists no longer had to contend with awkward and complicated schemes involving a mysterious ether. There is no unique frame of reference that marks an absolute state of rest. Otherwise, any object moving in that fixed box would be able to catch up to a light wave. That explained why Michelson and Morley detected no ether wind. The fixed ether had been a fiction all along.

There is no need to dwell on speeding spaceships to perceive a relativistic effect. Relativity can be measured right here on Earth. Cosmic rays from space crashing into the upper atmosphere create muon particles—heavy electrons—that spray downward at near the speed of light. But muons are extremely shortlived, lasting just millionths of a second, too little time for them to reach Earth's surface. But experiments show that they do make it to the ground. As relativity predicts, their inner clock appears to us to slow down, which extends their life just long enough to make it to the surface. From the muon's perspective, though, its lifetime is as short as it ever was; it's the distance between the upper atmosphere and the ground that has shortened, which allows the muon to make it to the ground.

All is relative, even mass. As an object approaches the speed of light, its mass increases noticeably, as measured by us. That's why nothing can go faster than the speed of light. Its mass would be infinite at that stage; no force exists that could push it faster, since the object would have infinite resistance. Einstein would later note that light itself has an effective mass. And since light is also energy, Einstein was able to link mass with energy in a universal law. His calculations showed the relationship to be $E = mc^2$, where c (as stated earlier) denotes the speed of light.

The teacher who once called Einstein a lazy dog, mathematician Hermann Minkowski, brilliantly cut to the quick and discerned an even deeper beauty in Einstein's new theory. ("I really wouldn't have thought Einstein capable of that," he remarked to a colleague about Einstein's accomplishment.) With his expert mathematical know-how, Minkowski recognized that he could recast special relativity into a geometric model. He showed that Einstein was essentially making time a fourth dimension. Space and time coalesce into an entity known as space-time. Time is the added dimension that allows us to follow the entire history of an event. You can think of space-time as a series of snapshots stacked together, tracing changes in space over the seconds, minutes, and hours. Only now the snapshots are melded together into an unbreakable whole. Dimensionally, time is no different than space. "Henceforth," said Minkowski in a famous 1908 lecture, "space by itself, and time by itself, are doomed to fade away into mere shadows, and only a kind of union of the two will preserve an independent reality."

Six years earlier Minkowski had moved from Zürich to become a professor at Göttingen. Although he had made a number of important contributions to number theory and other areas of pure mathematics, he is largely remembered for his reinterpretation of special relativity. It was easy for him to see that special relativity worked within a framework that had already been set up by mathematicians. "The physicists must now to some extent invent these concepts anew, laboriously carving a path for themselves across a jungle of obscurities, while very close by the mathematicians' highway, excellently laid out long ago, comfortably leads onwards," he said. To his mathematical eyes, special relativity was no more complicated than saying the world in space and time is a four-dimensional Riemannian manifold. To put it more plainly, Minkowski cleverly recognized that, while different observers in different situations may disagree on when and where an event occurred, they *will* agree on a combination of the two. From one position, an observer will measure a certain distance and time interval between two events. Perched in another frame of reference, a different observer may see more space or less time. But in both cases they will see that the *total* space-time separation is the same. The fundamental quantity becomes not space alone, or time alone, but rather a combination of all four dimensions at once—height, width, breadth, and time. Einstein, ever the physicist, was not impressed. When first acquainted with Minkowski's idea, he declared the abstract mathematical formulation "banal" and "a superfluous learnedness."

Oftentimes it is portrayed that the layman railed against the idea of

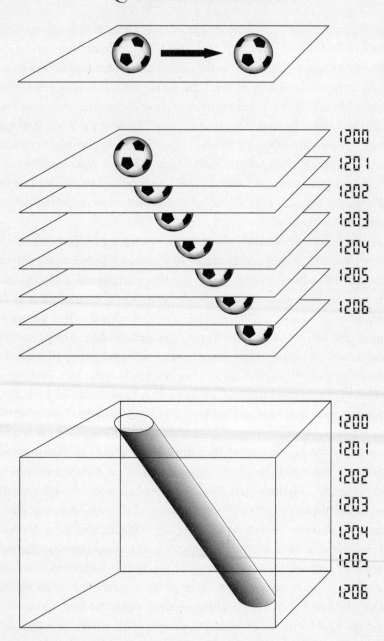

The simple movement of a soccer ball across a field is translated into space-time coordinates. With each tick of the clock, a snapshot captures the progression. Melded together, they form a tube that represents the entire motion in space-time.

relativity when it was first introduced, while the scientist greeted it with open arms. But for many scientists of the time, especially those deeply invested in classical physics of the nineteenth century, it was a psychological shock. Of course, opportunities to check the edicts of relativity were few and far between at first. It was only after several decades had passed and technologies had advanced that seeing its effects became more commonplace. Some, though, would not accept special relativity on aesthetic grounds. William Magie, a professor of physics at Princeton University, stated in an address before the esteemed American Physical Society in 1911 that "the abandonment of the hypothesis of an ether at the present time is a great and serious retrograde step in the development of speculative physics. . . . A description of phenomena in terms of four dimensions in space would be unsatisfactory to me as an explanation, because by no stretch of my imagination can I make myself believe in the reality of a fourth dimension. . . . A solution to be really serviceable must be intelligible to everybody, to the common man as well as to the trained scholar. All previous physical theories have been thus intelligible. Can we venture to believe that the new space and time introduced by the principle of relativity are either thus intelligible now or will become so hereafter? A theory becomes intelligible when it is expressed in terms of the primary concepts of force, space and time, as they are understood by the whole race of man."

Critics were demanding that direct earthbound experience be the criterion of truth, rather than mathematical formulas. But they were shortsighted in believing that our earthly domain was the sole theater of experience. As the British astronomer Arthur Eddington noted in a lecture: "It has been left to Einstein to carry forward the revolution begun by Copernicus—to free our conception of nature from the terrestrial bias imported in it by the limitations of our earthbound experience." Before Copernicus, medieval scholars solemnly concluded that the Earth couldn't possibly be moving and turning. Otherwise, everything on the planet would be torn apart in the motion—clouds would get ripped out of the sky, and objects dropped toward a spinning Earth would obviously miss their mark because the Earth would have rotated around at great speed during the fall. Medieval thinkers had not yet mastered the concept of inertia, the tendency for objects to resist any change in their movement. (A falling object, already moving with the Earth's rotation, remains in sync as it drops.) When Copernicus placed the Sun at the cosmic hub, he thrust Earth into motion. He taught us to rethink our intuition based on new evidence. Einstein was doing the same.

Special relativity drew a line in the sand. On one side stood our past scientific history, when most physics theories could essentially be explained to the layperson. With a bit of hand waving and a reference to a mechanical model, a physical idea could be popularly illustrated. More important, the explanation did not violate the principles of common sense. But after 1905 the terrain suddenly changed. The world according to special relativity didn't seem to be describing our ordinary humdrum surroundings. Simple mechanical models no longer worked.

There is a reason we are fooled: we live in a rather privileged place. Temperatures are extremely low (compared to a star, for instance), velocities are far from warp drive, and gravitational forces are essentially weak—an environment where the effects of relativity are very, very small. No wonder relativity appears strange to us. But as some physicists have put it, we are not free to adjust the nature of space-time to suit our prejudices. We're perfectly happy to adjust to the fact that thunder—a sound wave—arrives later than the lightning flash. It's part of our normal experience. Harder to accept is the finite and constant speed of light. Light travels so fast—it can wrap around the Earth nearly eight times in one second—that everything appears to occur simultaneously here on terra firma. It's difficult to directly experience the fact that observers, separated by a certain distance, will disagree on the precise time an event occurred.

Special relativity was exactly that—special. It dealt only with a specific type of motion: objects moving at a constant velocity. Einstein was determined to extend its rules to all types of motion, things that are speeding up, slowing down, or changing direction. But special relativity was "child's play," said Einstein, compared to the development of a *general* theory of relativity, one that would cover these other dynamical situations, in particular gravity. He tried to incorporate gravity directly into his special theory for a 1907 review article, but he came to recognize that it could not be done so readily.

Over the ensuing years Einstein's reputation would grow and soar. He finally left the Swiss patent office in 1909 when he received his first academic appointment at the University of Zürich. Two years later he moved on to the German University in Prague. After a year he went back to Zürich as a professor at his old haunt, the Polytechnic, where he had been so undistinguished as a student. He attained the peak of professional recognition when, in 1914, he moved to the University of Berlin as a full professor and member of the Prussian Academy of Sciences. Over these many years he waged a mental battle, amid teaching responsibilities, a failed marriage, and World War I. He struggled with the problem of recasting Newton's laws of gravity in the light of relativity.

The first thing he recognized was that the forces we feel upon acceleration and the forces we feel when under the control of gravity are one and the same. In the jargon of physics, gravity and acceleration are "equivalent." There is no difference between being pulled down on the Earth by gravity or being pulled backward in an accelerating car. To arrive at this conclusion, Einstein imagined a windowless room far out in space, magically accelerated upward. Anyone in that room would find their feet pressed against the floor. In fact, without windows to serve as a check, you couldn't be sure you were in space. From the feel of your weight, you could as easily be standing quietly in a room on Earth. The Earth, with its gravitational field keeping you in place, and the magical space elevator are equivalent systems. Einstein reasoned that the fact that the laws of physics predict *exactly* the same behavior for objects in the accelerating room and in Earth's gravitational hold means that gravity and acceleration are, in some fashion, the same thing.

These thought experiments, which Einstein carried out liberally to get a handle on his questions, led to some interesting insights. Throw a ball outward in that accelerating elevator in space and the ball's path will appear to you to curve downward as the elevator moves upward. A light beam would behave in the same way. But since accel-

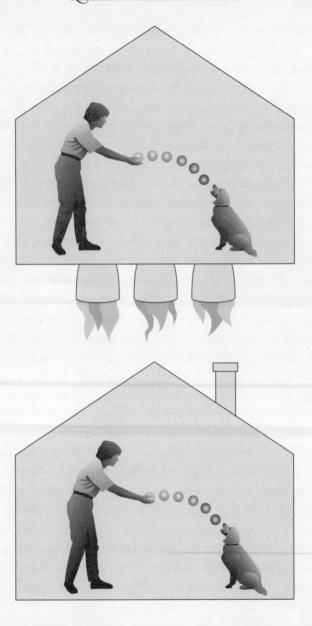

Thought experiment carried out by Einstein: a ball thrown in an accelerating room out in space falls toward the floor just as it does on Earth under the pull of gravity. Gravity and acceleration are equivalent. Einstein realized from this that a light beam should behave the same way, bending under the influence of gravity.

eration and gravity have identical effects, Einstein then realized that light should also be affected by gravity, being attracted (bent) when passing a massive gravitational body, such as the Sun.

Driven by his powerful physical intuition, Einstein began to pursue these ideas more earnestly around 1911 while he was in Prague. At that time he was beginning to confirm that clocks would slow down in gravitational fields (an effect never before contemplated by physicists). He was also coming to understand that his final equations would likely be "non-Euclidean." It was slowly dawning on him that gravity might involve curvatures of space-time. He was finally appreciating Minkowski's mathematical take on special relativity and its creation of space-time, that "banal" four-dimensional Riemannian manifold. Without Minkowski's earlier contribution, said Einstein contritely, the "general theory of relativity might have remained stuck in its diapers." Minkowski did not live to hear that; he had died in 1909 of appendicitis at the age of 44.

Returning to Zürich in August 1912, Einstein was eager to fashion his burgeoning conjectures into the proper mathematical format. Ignorant of non-Euclidean geometries, though, he joined up with mathematician Marcel Grossmann, an old college chum, to assist him in mastering the intricacies of this new mathematics. It was Grossmann who pointed out to Einstein that his ideas would best be expressed in the language of Riemann's geometry, by then advanced and extended by other geometers. In the spring of 1913 their collaboration generated a paper with all the essential elements of a general theory of relativity. As science historian John Norton would note, "Einstein and Grossmann had come within a hair's breadth of . . . the final theory." But they backed off from their findings. The two convinced themselves, based on some misconceptions, that their equations could not reproduce Newton's laws of gravity for the simplest cases. Newton's laws might be incomplete, but they were not wrong. They would still hold when gravity was weak and velocities were low. But unable to retrieve Newton under those simpler conditions, Einstein and Grossmann abandoned this line of attack, assuming it was the wrong choice. This misunderstanding, as well as the knowledge that their equations were not as yet completely universal, kept them from grasping success. For the equations to work, they still had to use a special reference frame, which meant they hadn't met the standard of developing a "general" theory. By April 1914, Einstein moved from Zürich to Berlin, which ended the collaboration with his friend. Einstein continued on his own, inexorably amending and tweaking his solutions, but now additionally armed with the mathematical insights introduced to him by Grossmann.

By the autumn of 1915, Einstein was becoming increasingly frustrated. His current theory, as it then stood, could not accurately account for a particular motion in the orbit of Mercury. Einstein was then predicting a shift of 18 arcseconds per century for this peculiar motion. He was aiming for the measured change of 45 arcseconds (measurements today peg it at 43). From his earliest days of contemplating a general theory of relativity, Einstein knew that a successful formulation of a new law of gravity would have to account for that anomaly.

The orbit of Mercury, a planet positioned about 36 million miles from the Sun, slowly revolves in the plane of the solar system. Imagine the orbit as an elongated ring. The point of the ring that is closest to the Sun—what is known as a planet's perihelion—shifts around over time. For Mercury the perihelion advances about 574 arcseconds each century.* Most of this shift is due to Mercury's interaction with the other planets; their combined gravitational tugging alters the orbit. But that can account for only 531 arcseconds. The remaining 43 arcseconds were left unexplained, a nagging mystery to astronomers for decades. Newton's laws couldn't resolve the discrepancy, at least given the known makeup of the solar system. That led some to speculate that Venus might be heavier than previously thought or that Mercury had a tiny moon. The most popular solution suggested that another planet, dubbed "Vulcan" for the Roman god of fire, was

---

*A circular orbit encompasses 360 degrees. There are 60 arcminutes in a degree and 60 arcseconds in each minute. Thus, 574 arcseconds is nearly 1/2,250 of an orbit. Mercury's orbital axis makes a complete revolution in roughly a quarter million years.

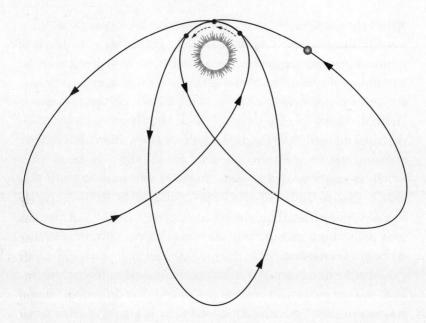

Over time, the point of Mercury's closest approach to the Sun—its perihelion—advances. The perihelion makes a complete turn around the Sun in 250,000 years. (The orbit's ellipticity is exaggerated for illustrative purposes.)

orbiting closer to the Sun than Mercury, providing an extra gravitational pull. There were even a few reports of Vulcan sightings, but none were reliable.

Then Einstein noticed a mistake in one step of the derivations he had conducted with Grossmann. This spurred him to consider that approach once again. He began to modify the equations and in the process became aware of his earlier misunderstandings. This allowed him to begin seeing that he could recover Newton's equations when gravitational fields were weak. His major effort took place during November 1915. On each of the four Thursdays of that month he reported his incremental progress to the Prussian Academy. A breakthrough came soon after his second report on November 11. That week he was at last able to successfully calculate the orbit of Mercury. He would later remark that he had palpitations of the heart upon seeing this result: "I was beside myself with ecstasy for days." It was the theory's first empirical success, grounding it in the real world. Moreover, Einstein's new formulation also predicted that starlight would get deflected around the Sun twice as much as he had earlier calculated (and twice the amount if Newton's theory is used). Triumph arrived on November 25, the day he presented his concluding paper entitled "The Field Equations of Gravitation." In this culminating talk he presented the final modifications to his theory, which no longer needed a special frame of reference. At last it was truly a *general* theory of relativity. In a letter to fellow physicist Arnold Sommerfeld, Einstein noted that he had just experienced "one of the most exciting, most strenuous times of my life, also one of the most rewarding."

What he discovered by working within his new universal framework was the very origin of gravity. Written in the deceptively simple notation of tensor calculus, shorthand for a larger set of more complex equations, the general theory of relativity displays a mathematical elegance:

$$R_{\mu\nu} - \tfrac{1}{2}\, g_{\mu\nu}\, R = T_{\mu\nu}$$

On the left side of the equation are quantities that describe the gravitational field as a geometry of space-time. On the right side is a representation of mass-energy and how it is distributed. The equal sign sets up an intimate relationship between these two entities. The two are intertwined: matter becomes the generator of the geometry. Consequently, gravity is not a force in the usual sense. It is actually a response to the curvatures in space-time. Objects that appear to be manipulated by a force are just following the natural pathways along those curves. Light, as it gets bent, is following the twists and turns of the space-time highway. Mercury, being so close to the Sun, has more of a "dip" to contend with, which partly explains the extra shift in its orbit.

Space-time and mass-energy are the yin and the yang of the cosmos, each acting and reacting to the other. The very cause of gravity is

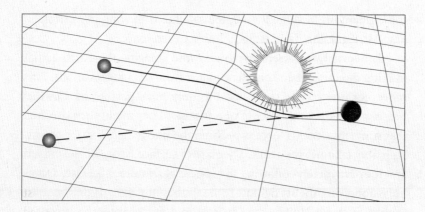

According to general relativity, space-time is like a vast rubber sheet. Masses, such as the Sun, indent this flexible mat, curving space-time. A star's light (solid line) traveling through the cosmos follows these space-time curves. Tracing the light back as a straight-line path (dashed line), it appears to us that the star has shifted its position in the celestial sky.

rooted in this image: it is the manifestation of the geometry of space-time. What Riemann suspected, Einstein firmly established. Einstein was not at all influenced by Riemann's vague yet prescient thoughts about a metrical field (Riemann never imagined the necessary ingredient called space-time), but he was greatly beholden to Riemann's mathematics. Space, Einstein taught us, may be thought of not as an enormous empty expanse but as a sort of boundless rubber sheet. Such a sheet can be manipulated in many ways: it can be stretched or squeezed; it can be straightened or bent; it can even be indented in spots. This image of space-time as a two-dimensional sheet is often used to help us visualize the concept, but the curvatures, of course, are imprinted on the full four dimensions of space and time. So, massive stars like our Sun are actually sitting in a flexible four-dimensional mat, creating deep depressions. Planets then circle the Sun, not because they are held by invisible lines of force, as Newton had us think, but because they are simply caught in the natural hollow carved out by the star. The more massive the object, the deeper the depression. Earth, for instance, is not holding on to an orbiting satellite with some phantom towline. Rather, the satellite is moving in a "straight" line—straight, that is, in its local frame of reference.

Think of two ancient explorers, who imagine the Earth as flat, walking directly north from the equator from separate locations. They move not one inch east or west but only push northward. But they hear they are moving closer to one another. They might then conclude that some mysterious force is pushing them together. A space traveler high above knows the truth. The Earth's surface is, of course, curved, and they are merely following the spherical contour. Likewise, a satellite is following the straightest route in the four-dimensional warp of space-time carved out by the Earth. As long as a heavenly body continues to exist, the indentations it creates in space-time will be part of the permanent landscape of the cosmos. What we think of as gravity—the tendency of two objects to be drawn toward each other—is a result of these indentations. Newton's empty box was suddenly gone. Space was no longer just an inert arena. Einstein showed us that space-time, the new physical quantity he introduced to physics, is a real-time *player* in the universe at large. Years later, reminiscing on this accomplishment, Einstein would write, "Newton, forgive me."

Starlight Waltz

*I*t was the distinctive elegance of his theory that gave Einstein confidence in its validity. "Its artistry resides in its inevitability, the economy of its structure, the basic simplicity that shines through its complexities, and a pervasive beauty that, like all beauty, defies analysis," said Banesh Hoffmann, who once collaborated with Einstein. In 1930 Einstein himself wrote that he did "not consider the main significance of the general theory of relativity to be the prediction of some tiny observable effects, but rather the simplicity of its foundation and its consistency." And yet it would be one of those "tiny observable effects" that turned Einstein into a living legend.

Of course, the anomalous precession of Mercury was already known, and general relativity could explain it. But Einstein also predicted another effect only pondered by Newton but never pursued. Once physicists thought of light as a wave, they generally assumed that it was different from matter, that it was immune from gravitational attractions.* General relativity, on the other hand, asserted that light *must* bend—that is, be attracted—by massive objects like the Sun, just like matter. More than that, the attraction would be twice the bending calculated when Newton's law of gravity is used. The extra contribution comes from the warping of space-time, especially near the mass, an effect that Newton's theory couldn't account for at all. Observing how a star's light got deflected around the Sun was thus a way of detecting the curvatures—the sloping space-time valleys—that Einstein was proposing in his theory. Of course, it's not the light that is actually bending, although that is how it is commonly described. More correctly, it is light's space-time path being flexed.

Near the Sun, a relatively lightweight star, the effect is very, very tiny. Einstein calculated that a ray of starlight just grazing the Sun's surface should get deflected by a mere 1.7 arcseconds (about 1/2,000 of a degree). That's roughly the width of the lead in a pencil seen from a football-field's length away. The bending diminishes the farther the light beam is situated away from the Sun and out of the solar space-time valley. In the spring of 1919, shortly after the close of World War I, Arthur Eddington, the British astronomer noted for his work on stellar physics, led a government-sponsored expedition to the tiny Isle of Principe, off the coast of West Africa, to look for this miniscule light deflection during a solar eclipse. An eclipse offered the perfect opportunity to view a star near the Sun as the Moon blocked its dazzling glare. Luckily, the eclipse was occurring in a part of the celestial sky with an exceptional patch of bright stars. To minimize the risk of bad weather, another team of astronomers journeyed to the village of Sobral in northern Brazil.

On the fateful day, May 29, Eddington and his teammate took 16 photographs, most of them ultimately useless because of intervening clouds. "We have no time to snatch a glance at [the Sun]," wrote Eddington of his adventure. "We are conscious only of the weird half-light of the landscape and the hush of nature, broken by the calls of the observers,

---

*Before Einstein, there were a few sporadic attempts to consider the effect of gravity on light. In 1804 J. Soldner, a German, published a short paper predicting the deflection of starlight by the Sun by an amount that follows from Newtonian theory, half that predicted by general relativity.

and beat of the metronome ticking out the 302 seconds of totality." In the end, two pictures did turn out to have good images of the essential stars. Within a few days, as a precaution against any mishaps on the voyage back, one of the plates was examined on the spot. Eddington and his companion carefully compared the picture to another photo of the same celestial region, taken months earlier in England when the Sun was not in the way. Eddington, who freely admitted he was unscientifically rooting for Einstein, was elated to see that the stars near the Sun had indeed shifted their apparent positions and by an amount that matched Einstein's prediction, give or take 20 to 30 percent. For Eddington that was close enough. It was certainly a larger bending than one would get using Newton's laws alone to calculate light's gravitational attraction. Here was proof, marginal as it was, that the long-reigning king of gravity, Newton, had been overthrown. Eddington would later remark that this was the most exciting moment in his life as an astronomer.

The Sobral expedition, which had fine weather and so was able to take many more photographs, confirmed the verdict. Einstein, ever confident, never doubted that the light deflection would be verified but was pleased, nonetheless, when he was informed via the scientific grapevine. He quickly dashed off a postcard to his mother to tell her the good news. The Royal Astronomical Society and the Royal Society of London, a scientific organization that Newton himself once presided over, held a special joint meeting that fall to officially announce the results, an example of the universality of science. A devastating war had just ended between Germany and Great Britain, and yet the British scientists were honoring a theoretical achievement made in enemy territory.

With its front-page reports of the solar eclipse experiment, the press on both sides of the Atlantic turned the name Einstein into a synonym for genius. His life in public was never the same again. Over the years, celebrities, from presidents to movie stars, clamored to wine and dine him. He was besieged with autograph requests. Photographers and artists regularly arrived at his doorstep to do his portrait. Even Cole Porter included the acclaimed physicist's name in a 1943 song entitled *It's Just Yours*: "Your charm is not that of Circe's with her swine/Your brain would never deflate the great Einstein." To this day his bushy mustache, helter-skelter hair, and world-weary eyes are instantly recognizable and remain an icon in cartoons and advertisements. "I have become rather like King Midas, except that everything turns not into gold but into a circus," remarked Einstein on his superstardom. For a man of thought, who yearned for a life of quiet contemplation, it was a state of affairs that he deemed "a dazzling misery."

Einstein died in 1955 and so did not live to see the further experimental triumphs of his theory in the latter half of the twentieth century. With the entrance of new astronomical techniques, light deflection experiments could be performed with much finer care than Einstein ever dreamed of. Solar eclipse experiments were carried out nine more times between 1922 and 1973, yet with only modest improvements in accuracy. Far better have been observations using a network of radio telescopes electronically linked around the globe. In this way one huge radio telescope as big as the Earth is created. Those who continued to question the validity of the coarse solar eclipse measurements were at last satisfied. By using this globe-spanning radio network to observe distant quasars, extremely intense and compact sources of radio waves, radio astronomers have been able to monitor how the apparent separation between close pairs of quasars changes as their radio signals pass close to the Sun. The accuracy in this type of measurement is nearly a thousand times better than Eddington's first crude try.

One of the most recent light deflection checks was a space-age version of the 1919 test but without the solar eclipse. The Hipparcos satellite, launched by the European Space Agency in 1989, spent four years making the most accurate measurements of stellar positions ever assembled. It did this for stars down to a magnitude of 10 (roughly 1,500 times fainter than the stars in the Big Dipper). The result: Einstein's prediction continues to hold up and with near perfection. In fact, Hipparcos's data were so precise that the Sun's ability to bend starlight could be detected halfway across the celestial sky. Stars located far from the Sun on the celestial sphere were observed to experience a shift in their apparent position, though far more weakly than for stars positioned closer to the Sun.

In 1964 Harvard-Smithsonian astronomer Irwin Shapiro, then with MIT's Lincoln Laboratory, came up with an entirely new and interesting variation to general relativity's light-bending effects.

Shapiro suggested transmitting a radar pulse from Earth and reflecting it off another planet. This technique had already been used to measure distances to nearby planets. But if the pulse passed by the edge of the Sun, Shapiro figured the radar signal's excursion to and from the planet should take a bit longer than what it would take were the Sun not there. That's because the Sun's warp of space-time, in a sense, adds a tad more distance to the journey; the radar beam must "dip into" the depression. Within two years the test was carried out. Radar signals were transmitted to both Venus and Mercury, as the planets were about to pass behind the Sun. For Venus the round-trip took about 30 minutes. Three hundred kilowatts of power were sent from a radar transmitter at the Haystack observatory in northeastern Massachusetts; the echo that returned was as small as 10^{-21} watt. But that was enough to notice that it took the signal about 1/5,000 second longer to return to Earth when the signal passed near the Sun. The path had lengthened by nearly 40 miles. Later, signals from the Viking landers on Mars, which set down in 1976, also were found to be delayed when passing near the Sun. Agreement between the measurements and the delay predicted by general relativity was within 0.1 percent, one part in a thousand.

The most beautiful example of light bending in the universe is gravitational lensing. Take the case of Abell 2218, a compact and rich cluster of galaxies situated more than a billion light-years from Earth. The view is breathtaking. Several bulbous elliptical galaxies sit like contented Buddhas in the middle of Abell 2218. A number of bright disks—spiral galaxies most likely—surround them. But there's more. Wispy arcs, 120 in all, encircle the entire heart of the cluster. The streaks are arranged like the rings of a dartboard. It is one of the universe's most wondrous illusions, created when Einsteinian light deflection is taken to the extreme.

When starlight passes by the Sun and gets bent, or deflected, the Sun is really acting like a lens. Recall that when you look through an optical lens it allows the object behind it to be magnified and brightened. It's a simple magnifying glass. A gravitational lens acts in a similar manner, only now it is gravity doing the work rather than a curved piece of glass. Soon after Eddington's successful solar eclipse test, Einstein and others discussed the possibility of light deflections—lensing—occurring farther out in space, as light passed by faraway stars. Depending on the orientation of the "lens," objects behind it could be simply magnified or split into multiple images, as if some giant fun-house mirror were at work. But by 1936 Einstein concluded that "there is no great chance of observing this phenomenon" beyond the Sun, since the chances for two stars being properly aligned were too small. Caltech astronomer Fritz Zwicky, though, had a grander vision. In 1937 he declared that galaxies offered "a much better chance than stars for the observation of gravitational lens effects." Zwicky was right, although it would take four more decades before his visionary insight was confirmed. The first such cosmic lens was sighted in 1979 (totally by accident). Since then dozens of lenses have been found. Some are single galaxies; others are entire clusters of galaxies, like Abell 2218. The cluster, trillions of times more massive than a single star like our Sun, collectively acts like a monstrous spyglass, greatly brightening the objects that lie far behind it. The faint blue arcs that surround Abell 2218 are actually the distorted ghostlike images of distant galaxies that reside some 5 to 10 times farther out. This makes gravitational lensing more than a cosmic curiosity. As witnessed by Abell 2218, lenses can act as a giant zoom lens. They take distant galaxies too faint to be seen and bring them into view. In this way astronomers manage a peek at the universe when it was far younger. No wonder lensing has been called "Einstein's gift to astronomy." Awareness of lensing effects is actually becoming quite vital to astronomers. Otherwise, it can lead to some astronomical bloopers. When galaxy FSC 10214+4724 was first discovered in 1991, for example, it was heralded as the most luminous galaxy in the universe. Though bright, it's not that brilliant. The Keck telescope in Hawaii later revealed that this galaxy is being brightened by a gravitational lens, a foreground galaxy located closer in. Oops, fooled by a gravitational illusion!

When Einstein first proposed his theory of general relativity in 1915, he made another prediction that could not be detected as readily as light deflection. Scientists then had neither the instruments nor the techniques to measure this extremely tiny effect. It was Einstein's declaration that time will pass more slowly in a gravitational field. To put it another way, a clock in space will tick more quickly than one "weighed down" by Earth's gravity. This situation is best imagined when we think of the gravitational force—the way Einstein first did—as equivalent to the force one feels in an accelerating room out in space. Picture a clock on the floor of that space elevator, with you on the ceiling observing it. But the room is accelerating upward. By the time the clock's ticks (marked by pulses of light) reach you at the ceiling, you and the ceiling have moved away in the motion upward. As the elevator moves faster and faster to mimic gravity, the peaks of the light waves will arrive at the ceiling at a slower rate. (In other words, the frequency will decrease.) Thus, the clock to you appears to be slowing down.

But, as Einstein taught us, the force experienced in this accelerating elevator is exactly mimicking the gravitational force on Earth. Hence, a clock on Earth would also tick away slower than one freely floating in space. This was a prediction unanticipated by any other physical theory. It was entirely new to physics. We don't notice this effect ourselves, for the atoms in our body are slowing down as well. We would know only by comparison. For example, any person who could miraculously survive on the surface of a neutron star, whose gravitational field is a trillion times stronger than Earth's, would age noticeably slower than a person more loosely grounded on terra firma. While a decade passes by on Earth, Neutronians would experience around eight years. Black holes, the mightiest gravitational sinkholes in the universe, carry this effect to the extreme. When a fraction of a second ticks away near the edge of a hole, many eons pass by in the rest of the universe. Relativity, in this case, lives up to its name. During a conversation with Einstein, writer Ashley Montagu once regaled the physicist with a popular joke concerning this paradox. It involved two men from the Bronx:

"What is relativity?" asks the first man.

The second man replies: "Supposing an old lady sits in your lap for a minute, a minute seems like an hour. But if a beautiful girl sits in your lap for an hour, an hour seems like a minute."

"And this is relativity?" responds the first.

"Yes," answers his companion. "That's relativity."

"And from this Einstein earns a living?"

There's another way to look at this effect. You might think of light waves as springs—coils that get stretched as they attempt to climb out of the "gravitational well" dug into space-time by a massive celestial object. Shorter waves, such as blue or yellow light, would thus get longer as they soar upward, shifted toward the other end of the electromagnetic spectrum. They would get redder. Hence, the name for this effect—gravitational redshift. The reddening is so miniscule in the neighborhood of the Earth and Moon that scientists had to wait until 1959 to measure it. Robert Pound, along with his student Glen Rebka, detected the redshift by setting up an experiment on the campus of Harvard University. They measured how gamma rays shifted their frequency ever so slightly as the energetic waves either ascended or descended a 74-foot-high tower within the Jefferson Physical Laboratory. The gamma rays came from a source of radioactive iron. To reduce the chance of the gamma rays getting absorbed by the dense air, a long Mylar bag was run through the tower and filled with lighter helium. The frequency changed within 10 percent of what Einstein predicted. Five years later, Pound and his colleague Joseph Snider got it down to 1 percent.

By the 1970s this gravitational redshift was being measured to levels of astounding accuracy. Atomic clock builder Robert Vessot of the Harvard-Smithsonian Center for Astrophysics rocketed into space one of his extremely accurate timepieces, a hydrogen maser clock, and compared its frequency to a similar clock on the ground. This special 90-pound clock, so regular and exact that its time varied by about a billionth of a second each day (that's roughly equivalent to one second every three million years), was launched aboard a four-stage Scout D rocket from Wallops Island off the eastern shore of Virginia on June 18, 1976. The launch occurred near dawn. Impact was 118 minutes later in the mid-Atlantic Ocean, 1,000 miles east of Bermuda. There was one nervous moment when Vessot and his colleagues lost contact with the spacecraft, but a minute later the signal was reacquired. A circuit breaker had accidentally cut off the power supply to the uplink transmitter. The basic idea of the experiment was simple—monitor the oscillations of the atomic clock as it traveled almost vertically up some 6,000 miles and then down again. In the end they found that at

6,000 miles above the Earth, where gravity has loosened its grip, the clock did indeed run a bit faster, by some 4.5 parts in 10^{10}. If it had been there in orbit for 73 years, it would have gained a whole second compared to a clock on Earth. The accuracy of the test was within a hundredth of a percent, 100 times better than the gravitational redshift measurements on the Harvard campus.

In 1976 such a test had little practical importance, but that's no longer true. The high-stability clocks aboard the Global Positioning System satellites, perched high above Earth, are regularly affected by the gravitational redshift. Twenty-four in all around the globe, these satellites must be synchronized to within 50 billionths of a second to allow users to know their position on the ground to 15 yards. But without a relativistic correction, the clocks would run faster by 40,000 billionths of a second each day, most of that due to the gravitational redshift. Periodic corrections are programmed in, otherwise the clocks would be out of sync within a minute and a half. Clifford Will knew that general relativity had finally arrived when he had to prepare a briefing on the theory for an Air Force general, as it became a matter of national security that the Global Positioning System be as accurate as possible. Hollywood recognized the drama of this situation in the James Bond movie *Tomorrow Never Dies*, where an evil genius attempts to insert errors into the system to send British ships into harm's way.

The gravitational redshift was not the only novel and strange effect predicted by general relativity. In a 1913 letter to the Austrian physicist and philosopher Ernst Mach, Einstein mentioned a new force that should come into play with general relativity. He called it "dragging." This was two years before Einstein had worked out his full theory. In many ways, dragging is to gravity what magnetism is to electricity. In fact, some call it *gravitomagnetism*. A charged particle as it spins creates a magnetic field that surrounds the particle; similarly, a spinning mass, such as the turning Earth, imparts a rotation to the surrounding medium, which is space-time itself! In 1918 two Austrian scientists, Josef Lense and Hans Thirring, calculated the effect such a spin would have. Consequently, it is sometimes called the Lense-Thirring effect.

Lense and Thirring saw that an object spinning pulls the very

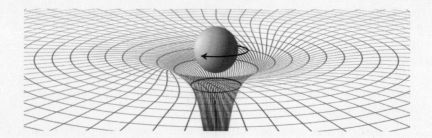

Frame Dragging: As a black hole spins, it twists space-time around itself.

framework of space-time around with it, like a cake batter swirling around the beaters in an electric mixer. The whirling is strongest nearest the beaters and gradually diminishes farther away. In 1959 a magazine ad in *Physics Today* for a new kind of gyroscope sparked some physicists (while swimming naked in the Stanford University pool and musing as they exercised) to imagine the perfect gyroscope and how it could be used to measure this subtle feature of general relativity. By 1963 they obtained support from the National Aeronautics and Space Administration (NASA). For more than 30 years, in fits and starts and rising like a phoenix to survive seven cancellations, the project continues. At a cost of $500 million, the endeavor is highly controversial. Called Gravity Probe B (Vessot's experiment was Gravity Probe A), it is one of the most expensive pure science projects that NASA has ever sponsored (and the longest in preparation).

The plan is to use a set of four gyroscopes and launch them 400 miles above the Earth in polar orbit. A gyroscope is essentially a spinning wheel. For the Gravity Probe B, the spinning gyroscopes will be four quartz globes, each a mere 1½ inches in diameter. The globes are coated with a layer of niobium, giving them a silvery finish. They could be registered in the *Guinness Book of World Records* as the smoothest, roundest objects on Earth. They have been polished to within 50 atomic layers of being perfect spheres. Such perfection is needed to measure the tiny changes at work; distortions could introduce a mechanical wobble that might be mistaken for space dragging. Once set in motion and free of outside disturbances, the axes of these spinning globes should keep pointing in the same direction. Because of conservation of angular momentum, the globes will resist any change in their orientation. This makes them the perfect tool to measure space-time dragging. Gyroscopes spinning in space can be aligned with a far-off star. But over time, as the spinning Earth drags local space-time around itself, this alignment will slowly drift. Each gyroscope's axis, while maintaining its direction with respect to local space-time, will no longer align on the far-off star. This wobble is not a large effect. The image of the swirling batter is far too strong when it comes to Earth, a lightweight celestial object. According to general relativity, the axis of a spaceborne gyroscope should move an infinitesimal 0.000012 of a degree each year because of Earth's dragging the framework of space-time around itself. That's the width of a human hair seen from a quarter mile away.

While this effect is virtually meaningless to Earth's cosmic life, such "frame dragging" may have far bigger consequences in other environments, such as in quasars. A powerful young galaxy caught at the edge of the visible universe, a quasar emits the light of tens of normal galaxies, with most of that energy believed to be generated by a supermassive black hole in the quasar's center. The black hole contains the mass of hundreds of millions of suns. With such a large mass spinning, the magnitude of frame dragging is gargantuan. In fact, some speculate that it causes any nearby matter to spiral in toward the black hole's poles, which then shoots straight outward in spectacular jets that span hundreds of thousands of light-years. In this situation, frame dragging is made highly visible.

The noted Danish physicist Niels Bohr, who first conceived of an atom's inner structure, visited the United States in January 1939 to work a few months with Einstein at the Institute for Advanced Study in Princeton, New Jersey. But right before his ship, the MS *Drottningholm*, left Europe, he learned that nuclear fission had been discovered. German scientists had verified that their uranium nuclei were splitting into smaller pieces. Arriving at Princeton greatly excited by the news, Bohr immediately began working on the problem with John Archibald Wheeler, then 27, the Princeton physics department's newest addition and a specialist in atomic and nuclear physics. Together, they developed a general theory of nuclear fission. Using it they predicted that such nuclei as uranium-235 would be effective in sustaining chain fission reactions. Wheeler went on to become a central figure in subsequent historic developments in physics, including World War II's Manhattan Project and the later development of the hydrogen bomb. Upon returning to Princeton in the early 1950s after this war work, though, he chose to move in a completely different direction. "I suppose it was infection," says Wheeler. "As a student I had read a book called *Problems of Modern Physics* by H. A. Lorentz, a great father figure in physics. And what were the problems? They were quantum physics and relativity." Having spent years on the first problem, Wheeler decided to tackle the second. It was a dicey decision. Relativity had turned into a backwater in physics, a field inhabited by lone specialists. "There were all these people working with Einstein who didn't know the rest of physics," recalls Wheeler.

For several decades general relativity had been the most admired yet least verified theory in physics. There was the subtle twist of a planetary orbit here, the tiny bending of a beam of starlight there. The theory could also account for the expanding universe discovered by Edwin Hubble in the 1920s. Yet even with that, the experimental evidence was admittedly thin. Not until midcentury did things begin to change, largely due to the new technologies that could better assess the minute changes predicted by relativity. By the 1960s general relativists were entering a golden age of experimentation. Pound and Rebka at last measured the gravitational redshift, while Shapiro came up with an intriguing new way to measure space-time curvatures. But this renaissance would not have occurred without one other vital factor: a concerted effort by theorists to study general relativity more deeply as well. And at the forefront of this movement to bring Einstein's theory back into the thick of mainstream physics and connect it to the universe at large was Wheeler. Almost single-handedly he would change general relativity's moribund image. He immersed himself in the subject through teaching. "Much of the best teaching comes out of research, and much of the best research comes out of teaching," notes Wheeler. "If the class hour doesn't end with the teacher having learned something, he doesn't know how to teach." It was then that Wheeler encapsulated general relativity in one clear sentence: "Mass tells space-time how to curve, and space-time tells mass how to move."

Einstein gave the last seminar of his life in Wheeler's class, which met in the physics department's former quarters, the Palmer Physical Laboratory. It's an impressive gothic-style building, constructed of red brick with a thick slate roof. Erected in 1907, the building is now used as a center for Asian studies, so the physics themes played out in the stained glass windows and statuary of honored physicists of the past are oddly out of place. Wheeler slowly walks a visitor through his early years in physics. He proceeds past the heavy wooden doors at the entrance and up the wide central staircase. On the second floor, after a turn, the first room on the left is number 309. Here is where Einstein gave his last classroom lecture. The chairs, in dark wood, have widened arms on the right for note taking. Real blackboards, the old-fashioned kind, completely line the walls at the front and along the right side of the room. The seats are eight across and eight deep. The room has the smell of old wood and chalkdust, rather nostalgic and not unpleasant. One can almost picture the elderly Einstein at the front, in his casual attire, stepping the students through his thoughts. Wheeler recalls Einstein talking about three things: first, how he came to relativity; second, what relativity meant to him; and, third, why he didn't like quantum theory, whose edicts went against his scientific philosophy. The role of the observer is central to quantum theory; nothing is known until it is measured by an observer. Wheeler remembers Einstein wondering aloud, "If a being such as a mouse looks at the universe, does that change the state of the universe?" In such timeworn surroundings, Wheeler would revive general relativity, taking it from its minor position in physics to one of its most thriving fields.

It began when Wheeler looked at a problem almost forgotten. What happens to a star that is particularly heavy? What happens to it at its death? J. Robert Oppenheimer (who would later head the Manhattan Project that constructed the first atomic bomb) and his student Hartland Snyder published a paper on this very question in the September 1, 1939, issue of *Physical Review*. (Coincidentally, Bohr and

Wheeler had their paper on the theory of fission in the same issue.) They began with a star that has exhausted all its fuel. With the heat from its nuclear fires gone, the star's core becomes unable to support itself against the pull of its own gravity, and the stellar corpse begins to shrink. If this core is weightier than a certain mass, now believed to be two to three solar masses, Oppenheimer and Snyder confirmed that the core would not turn into a white dwarf star (which will be our own Sun's fate) nor even settle down as a tiny ball of neutrons. From general relativity they calculated that the star would continue to contract indefinitely. It would be crushed to a "singularity," a condition of zero volume and infinite density conceived earlier by the German astronomer Karl Schwarzschild in 1916 when he was working with Einstein's newly published equations. It was a place where all the current laws of physics break down. The last light waves to flee before the door is irrevocably shut get so extended by the enormous pull of gravity (from visible, to infrared, to radio and beyond) that the rays become invisible and the star vanishes from sight. What remains is a spherical region of space out of which nothing—no signal, not a glimmer of light or bit of matter—can escape. The ethereal boundary of this sphere has come to be known as the "event horizon." It is not a solid surface but rather a gravitational point of no return. Once you've stepped inside that invisible border, there would be no way out, only a sure plummet into the singular abyss at the center. Space-time around the collapsed core becomes so warped that the stellar remnant literally closes itself off from the rest of the universe. "Only its gravitational field persists," wrote Oppenheimer and Snyder in their historic paper. For Schwarzschild such a condition was an interesting mathematical solution to Einstein's equations; Oppenheimer and Snyder were arguing that it could be the actual fate of a massive star.

But in 1939 Oppenheimer and Snyder didn't consider all of the forces that might possibly prevent such a dire finale. Coming back to the problem in the 1950s, Wheeler wondered if pressure, the resisting power of a substance, might change the result. Perhaps the pressure of the star's material would prevent the ultimate collapse. Or maybe in its death throes an aging star throws off so much radiation and matter that gravitational collapse is averted, and it settles down as a white dwarf or neutron star. "I was looking for a way out," says Wheeler. Kip Thorne, Wheeler's graduate student in the early 1960s, today speculates that Wheeler's resistance to accepting the star's dark fate may also have been partly due to the idea originating with Oppenheimer. Wheeler, a political conservative, had reservations about Oppenheimer, who was long publicly challenged for his liberal beliefs. They had been on opposite sides during the first governmental debate on the need for a hydrogen bomb. "There was something about Oppenheimer's personality that did not appeal to me," confessed Wheeler in his autobiography. "He seemed to enjoy putting his own brilliance on display—showing off, to put it bluntly. . . . I always felt that I had to have my guard up."

After Oppenheimer had worked briefly on the problem of "continued gravitational contraction," as he called it, he inexplicably dropped it, never taking it up again. "He didn't recognize the importance of it," explains Thorne. "But Oppenheimer's work with Snyder is, in retrospect, remarkably complete and an accurate mathematical description of the collapse of a black hole. It was hard for people of that era to understand the paper because the things that were being smoked out of the mathematics were so different from any mental picture of how things should behave in the universe." Wheeler was so skeptical, in fact, that he hardly mentioned the existence of Oppenheimer's paper in his early work on general relativity. His attitude didn't appreciably change until 1962 when David Beckedorff, an undergraduate student then completing a senior thesis at Princeton, reexamined the Oppenheimer solution and recast it in a simpler form. "It was a real eye-opener for me," says Thorne, just then starting his graduate work with Wheeler. With other loopholes closed off as well, particularly due to the introduction of computers that could simulate the difficult physics of an imploding star, Wheeler was finally convinced that the star had to collapse. "Even if you put in the most powerful attempt to fight collapse, you can't prevent it," he says. "You always end up with a 'gravitationally completely collapsed object,'" as he was then awkwardly calling it.

When pulsars were discovered in 1967, and not yet understood, a conference was quickly set up at NASA's Goddard Institute for Space

Studies in New York City to discuss the possible suspects. Could they be red giant stars, white dwarfs, or neutron stars? Wheeler lectured that astronomers should consider the possibility that they were his gravitationally collapsed objects. "Well, after I used that phrase four or five times, somebody in the audience said, 'Why don't you call it a black hole?' So I adopted that," says Wheeler today (although some suspect Wheeler carefully crafted the term himself after years of thought). Whatever the origin, Wheeler used the term again in a scientific lecture several weeks later. The name became official. Black hole—a term so appropriate, as it is truly a pit dug into the fabric of space-time—went into the scientific lexicon. The catchy phrase caught the public's imagination (and caused blushes for a while in France where *trou noir* has obscene connotations).

While doing his graduate work at Princeton, Thorne observed this rebirth of general relativity firsthand. As experimentalists were testing their cherished theory in ways impossible to perform in the past, Thorne came to recognize that these scientists needed theorists like him to help them out. In general relativity it is not easy to decide what to measure. It's a slippery theory. Depending on the coordinates you use, you can come up with apparently different answers. When carrying out his radar experiments, Shapiro was continually challenged, forced to justify his calculations to others again and again. It is one of the reasons that experimental relativity took nearly a century to bloom and flourish. It's a difficult business to decide what you will measure, how you will measure it, and how to interpret the results. Many have stumbled along the way. Controversies have broken out over what is and what is not possible to observe. Partly to resolve these conflicts, theorists realized they had to construct a more comprehensive scaffolding that contained not just Einstein's theory but alternate theories of gravity as well. "Although Einstein's theory of general relativity is conceptually a simple theory, it's computationally complex," says Thorne. "If you want to identify what it is you are testing in any given experiment, you basically need a larger framework than relativity itself gives. You need some other possibilities. You will then have a large class of conceivable theories of gravity in which relativity is one.

These other theories then act as a foil against which to examine relativity." Observers could set up experiments that tested for certain differences between these various theories, to see if Einstein's held up. Clifford Will, who studied under Thorne at Caltech and is now at Washington University in St. Louis, and Kenneth Nordtvedt of Montana State University analyzed and categorized many of these alternative theories and even invented a few of their own. "Partly as strawmen," explains Will. "It was a way of motivating an experiment and interpreting the results." Others offered revised equations of gravity because they truly believed that general relativity needed to be amended for various theoretical reasons. All these theories, says Will, "forced general relativity to confront experiment as never before." The most famous alteration to Einstein's equations—and for a while its most serious challenger—was the Brans-Dicke theory.

Princeton University was the center of the renaissance in general relativity and not solely in the theoretical arena. While Wheeler wrestled with gravitationally collapsed objects on paper, Princeton physicist Robert Dicke was reinvigorating relativity on the experimental end. "They thought about things in very different ways," notes Thorne. "Wheeler was a dreamer, driven to a great extent by physical intuition wrapped up in philosophy. Dicke was a gadgeteer who had theoretical dreams as well. But his ideas were radically different from Wheeler's. Wheeler thought about the universe in terms of geometry; Dicke thought about it in terms of field theory."

Dicke was a very generous scientist. In 1965 he helped Arno Penzias and Robert Wilson figure out that a pesky noise in their radio telescope at Bell Laboratories in New Jersey was actually the fossil whisper of the Big Bang, a buzz that had been echoing through the corridors of the universe for some 15 billion years. Dicke was just getting ready to look for this cosmic microwave background himself. Penzias and Wilson later won the Nobel Prize in Physics for their accidental discovery, but Dicke was not bitter by this turn of events at all. He simply said he was "scooped."

Dicke, who died in 1997, was revered by an entire generation of physicists as the premier experimentalist of his time. "Dicke made experimental discoveries or elucidated theoretical principles that led to, among other things, the lock-in amplifier, the gas-cell atomic clock, the microwave radiometer, the laser, and the maser," Will has written. "He was involved, directly or indirectly, in so many discoveries for which Nobel Prizes were awarded, that many physicists regard it as a mystery (and some as a scandal) that he [never received] that honor." Trained in nuclear physics, Dicke began thinking about gravity around 1960 when he concluded that gravitational tests up to that point were woefully imprecise. One of his earliest ventures in this arena was to redo the famous experiments of Baron Roland von Eötvös of Hungary, who in 1889 and 1908 tested the equivalence of inertial mass (the aspect of a body that resists acceleration) and gravitational mass (the mass that feels a gravitational attraction) with exquisite precision. That mass is affected by these separate forces in the exact same way is the very cornerstone of both Newtonian physics and general relativity. It is the reason that different masses, both light and heavy, fall at the same rate when dropped from a tall height (say, from the Leaning Tower of Pisa). A heavier mass is more attracted to the Earth than a lighter one. Yet at the same time it has a greater resistance to the acceleration— enough resistance to slow its progress and match the speed of its lighter companion in the overall fall. Eötvös found this match to be exact to a few parts in a billion. Dicke and his co-workers got it down to a few parts in 100 billion, and later a Moscow team headed by Vladimir Braginsky made some improvements. In the mid-1990s a group led by Eric Adelberger at the University of Washington in Seattle reached the part-in-a-trillion level. They called their experiment "Eot-Wash," a pun on the good baron's name, which is pronounced "ut-vush."

In turning to these questions Dicke began thinking very deeply about the foundations of gravitational theory itself. He came to believe in what is known as "Mach's principle," named after Ernst Mach, who first voiced the concept decades earlier. Essentially, this principle states that the strength of gravity depends on the distribution of matter throughout the entire universe. If that is true, then gravity's forceful-ness should diminish as the universe expands and diffuses its cosmic density. Dicke estimated at the time that the change would be roughly one part in 20 billion with each passing year. Einstein's general relativity did not allow for this at all. Working with his graduate student Carl Brans, Dicke incorporated Mach's principle into an alternate theory of gravity by adding an extra term to Einstein's equations. As a result, the Brans-Dicke theory came up with slightly different numbers on certain gravitational measurements, such as the perihelion of Mercury. Einstein seemed to have gotten that right, but he had assumed that the Sun was fairly spherical. What if the Sun were more oblate, more squished down than people were aware of, perhaps due to a rapidly spinning core? If so, Einstein would be wrong and the Brans-Dicke theory more useful. To find out, Dicke set out to measure the Sun's oblateness to a finer degree than had ever before been done. Then a postdoctoral researcher (postdoc) at Princeton, Rai Weiss recalls Dicke disappearing for a few weeks when he first got this idea. One Monday he walked back into the office with a fat sheaf of rolled-up drawings, about 50 or 60. Dicke had stepped through the entire experiment in his mind. He had designed the telescope, the electronics, and all the optics, as well as the supporting structure. The two assistants assigned to build the apparatus eventually saw that Dicke had anticipated every correction, adjustments usually not discovered until an instrument is under construction. He had done it all with pure thought.

"We called it a very Dickesque experiment," says Kenneth Libbrecht, a former graduate student of Dicke's, "because all really subtle and clever experiments that had everything chopping back and forth were called that among my graduate student friends." Dicke's experiment involved gathering two differing signals in quick succession, which is the purpose of a lock-in amplifier. It automatically shifts a detection back and forth in synchronization. It will first take data of both a signal and its background and then measure just the background. The instrument is programmed to do this continuously, cycle upon cycle. In the end, by subtracting the background from the overall data, a weak signal can emerge from the noise. Dicke was a coinventor of the lock-in amplifier and founded a company, Princeton Applied Research, to build such devices commercially. "As graduate students, we'd sit around and buzz about what Dicke was worth," recalls Libbrecht. "The rumor was about $10 million." His sole splurge was a cabin in Maine, where he spent a month each summer. "Other than that, he never acted rich. He wore his frumpy old tennis shoes to work, like everybody else."

Dicke, along with H. Mark Goldenberg, measured the Sun's roundness in 1966 by placing a circular disk—an occulting disk—in front of the Sun's image in the telescope. What remained was the very edge or limb of the Sun. Photodetectors basically scanned this thin circle of light to discern any fattening at the Sun's equator. Though the Sun does bulge due to its rotation, Dicke and Goldenberg reported far more bulging than previously assumed, enough to favor the Brans-Dicke theory over general relativity. It looked like Einstein was about to be overthrown. The sheer possibility is what motivated many to continue carrying out the classic tests of general relativity—light deflection, time delays, gravitational redshifts—to higher and higher degrees of precision so that the differing predictions of Einstein and Brans-Dicke could be appraised more rigorously. Over time, though, the enhanced measurements were found to match the predictions of Einstein's general theory of relativity far more than the predictions from Brans-Dicke or any other alternate theory of gravity. Yet Dicke's oblateness result, which seemed to hold up (or at least was not yet refuted), was still a lone and nagging concern. To settle the controversy, Libbrecht redid the solar measurements. Dicke had looked at the Sun from Princeton, which often has clouds in the sky. In the summer of 1983 Libbrecht set up a shack atop Mount Wilson just north of Pasadena in sunny California. "As a graduate student I was thinking, 'Not only am I going to knock down general relativity, but I'm going to revolutionize solar physics at the same time.' Then the whole thing fell down like a house of cards," says Libbrecht. All the effects that Dicke had analyzed over those many years appreciably disappeared. Princeton's cloudy skies had likely introduced errors. The Sun's oblateness is actually quite small. "That was Bob's last hurrah," notes Libbrecht. "But that's why he was such a hero. When the data said his theory was wrong, then for him his theory was wrong. It was as simple as that. He soon retired after that."

Pas de Deux

Through the work of theorists and experimentalists such as John Wheeler and Robert Dicke, space-time joined an ever-growing cast of characters in the universe's cosmic drama. This new participant took on a definite role and personality. Space-time became the universe's flexible stage, a rubbery structure that stars, planets, and galaxies could bend and dent in intriguing ways. This new outlook had dramatic consequences. It meant that, when an object embedded in space-time gets moved or jostled, it can generate ripples in this pliable space-time fabric. Jiggle a mass to and fro and it will send out waves of gravitational energy, akin to the way a ball that is bounced on a trampoline sends vibrations across the canvas. These gravitational (or gravity) waves will uniformly radiate outward much like light waves. But while electromagnetic waves move through space, gravity waves are undulations in space-time itself. They alternately stretch and squeeze space—stretch and squeeze somewhat like the bellows of an accordion in play. And as these ripples encounter planets, stars, and other cosmic objects, they will not be stopped. Rather they will simply pass right on through, as they expand and contract all the space around them.

Anything in the universe that has mass is capable of sending out gravity waves—all it has to do is move. But the strength of the signal depends on the amount of mass and the nature of its movement. A mammoth body like a star has a powerful gravitational pull, but since it remains essentially stationary (aside from its steadfast motion within the galaxy), it emits little gravitational radiation. Earth also continually emits weak gravitational energy as it circles the Sun, although it would take the age of the universe before we'd notice any effects from the emission. The Moon sends out still weaker waves as it moves around the Earth. Even hopscotch players have an infinitesimal chance of emitting a gravity wave or two as they jump up and down. The strongest waves, though, emanate from the most violent and abrupt motions the universe has to offer: stars crashing into one another, supernovas erupting, and black holes forming. Some of these events are not seen directly with electromagnetic radiation; thus, gravitational radiation offers a new means of exploring the universe. Gravity waves will not just extend our eyesight, they will provide an entirely new sense. "Gravitational waves could prove to be the most penetrating waves in nature. That is in part their charm but also their curse, since it makes them so difficult to detect," Rainer Weiss and LIGO director Barry Barish have written.

Einstein first discussed the concept of gravitational radiation in 1916, shortly after he introduced his theory of general relativity.* His paper on the topic was tucked away in the *Sitzungsberichte der Königlich Preussichen Akademie der Wissenschaften* (Proceedings of the Royal Prussian Academy of Science) next to articles on the perception of light by plants and the use of first person in Turkish grammar. An algebra mistake led Einstein to a misconception about the origin of

---

*As early as 1908, Henri Poincaré did mention that a relativistic theory of gravitation, yet to be established, would likely involve the emission of gravitational waves ("*onde d'accélération*").

gravity waves in this initial paper, but he made the correction in a follow-up paper in 1918. He recognized that just as electromagnetic waves, such as radio waves, are generated when electrical charges travel up and down an antenna, waves of gravitational radiation are produced when masses move about. Moreover, they would also travel at the speed of light. To picture the generation of a gravity wave, Einstein imagined a cylindrical rod spinning around, like a game of spin the bottle. In this case the frequency of the gravity wave emission would be twice that of the rotation. These waves would flow smoothly outward from the source, but because gravitational energy moving through space would disperse and grow weaker, in the same way starlight does, Einstein doubted that gravity waves would ever be observed, even from the most violent astronomical sources. By the time the gravity waves from an exploding star in our galaxy strike Earth, for instance, they are little more than subatomic flutters. Were a gravity wave from a supernova in the center of the Milky Way to hit this page, it would be so weak that it would squeeze and stretch the sheet's dimensions by a mere hundred-thousandth of a trillionth of an inch—a measure 10,000 times smaller than the size of an atomic nucleus.

Given the extreme weakness of the signal, few scientists were interested in the phenomenon when Einstein first described it. Why bother with an effect too small to detect? Furthermore, a lively debate ensued for some four decades on whether gravity waves existed at all. Many seriously wondered whether they were just artifacts—unreal ghostly products—in the equations of relativity. That possibility inspired Arthur Eddington to mischievously ponder whether the waves really "traveled at the speed of thought." Even Einstein had doubts at one point while he was working at the Institute for Advanced Study. These suspicions lingered into the 1960s. The doubts initially arose because there is a pitfall in general relativity: its equations are written in such a way that they are independent of all coordinate systems. So when a theorist dives in and chooses a particular system of measurement, the results can be tricky to interpret. For example, if your measurement system happens to have its coordinates fastened to the masses, a gravity wave passing by wouldn't budge them off the coordinates (within your calculations, that is), leaving the impression that the wave had no effect on space-time whatsoever. "A few years ago here at Syracuse University," recalls gravitational wave physicist Peter Saulson, "the department received a paper that seemed to prove that gravity waves would get absorbed by the interstellar medium. But it took the physicists here at the time several months to figure out exactly where the author of this paper went wrong. He had chosen a coordinate system unfamiliar to them. Each generation seems to have to work this out."

The controversy today is pretty well laid to rest to everyone's satisfaction. No one really questions anymore whether gravity waves truly exist. With all the evidence amassed in favor of Einstein's view of gravity, physicists are convinced that gravitational radiation is a natural consequence of the theory. This confidence, though, is not based on faith alone. Indirect yet exquisite evidence that gravity waves are real arrived in the 1970s, when radio astronomers uncovered one of nature's most dependable gravity wave emitters in the celestial sky. Their tale of discovery is composed of one part ingenuity, one part serendipity, and two parts sheer pigheadedness.

Only a month after finishing his Ph.D. in radio astronomy at Harvard University in 1967, Joseph Taylor heard about the discovery of a strange new object in the heavens. "This was a time when the journals were always publishing something quite new, but this was more unexpected than anything I can remember at the time," recalls Taylor. The discovery had been made using a sprawling radio telescope—more than two thousand dipole antennas lined up like rows of corn—near Cambridge University in Great Britain. Jocelyn Bell (now Burnell), then a Cambridge graduate student, was one of the laborers. "I like to say that I got my thesis with sledgehammering," she has joked. The telescope was designed by Cambridge radio astronomer Antony Hewish to search for quasars, and it was Bell's job to analyze its river of data. At a radio astronomy conference in 1983, Bell Burnell recalled the moment when she realized that some of the squiggles recorded on her reams of strip-chart paper didn't look quite right:

We had a hundred feet of chart paper every day, seven days a week, and I operated it for six months, which meant that I was personally responsible for quite a few miles of chart recording.

It was four hundred feet of chart paper before you got back to the same bit of sky, and I thought—having had all these marvelous lectures as a kid about the scientific method—that this was the ideal way to do science. With that quantity of data, no way are you going to remember what happened four hundred feet ago. You're going to come to each patch of sky absolutely fresh, and record it in a totally unbiased way. But actually, one underestimates the human brain. On a quarter inch of those four hundred feet, there was a little bit of what I call "scruff," which didn't look exactly like [man-made] interference and didn't look exactly like [quasar] scintillation. . . . After a while I began to remember that I had seen some of this unclassifiable scruff before, and what's more, I had seen it from the same patch of sky.

The 81.5-megahertz radio signal was emanating from a spot midway between the stars Vega and Altair. A higher-speed recording revealed that the signal was actually a precise succession of pulses spaced 1.3 seconds apart. The unprecedented clocklike beeps caused Hewish and his group to label the source LGM for "Little Green Men." This was done only half in jest. At one point some consideration was given to the possibility that the regular pulsations were coming from a beacon set up by an extraterrestrial civilization. Within a month Bell ferreted out, from the yards upon yards of strip chart that were spewing from the telescope, the telltale markings of a second suspicious source. Its period was 1.19 seconds. By the beginning of 1968, two more were uncovered. When the phenomenon was announced to the public, a British journalist dubbed the freakish sources pulsars.

Remaining at Harvard as a postdoctoral fellow, Taylor quickly rounded up a team to observe the four pulsars with the imposing 300-foot-wide radio telescope at the National Radio Astronomy Observatory in Green Bank, West Virginia (a dish that dramatically collapsed in 1986). While the original pulsars had been found by visually searching for specific peaks—pulses—on a paper chart recorder, Taylor developed a different strategy to look for more. A pulsar, of course, *beep*s with a regular beat, but it also has a sort of echo. As the pulsar signal travels through the thin plasma of interstellar space, its various frequencies propagate at varying speeds: the high radio frequencies travel faster than the lower frequencies. (The speed of light can vary when *not* in a vacuum.) Consequently, like horses on a racetrack, the differing radio waves start spreading apart. They get dispersed. By the time they reach Earth, the high-frequency pulse arrives first, followed by the lower frequencies in quick succession. Overall, the pulse appears extended, sweeping rapidly downward in frequency. Taylor and his colleagues wrote a special program to look for this distinctive profile in their streams of celestial radio data, automating the search with the use of a computer. "No one had thought about doing this kind of thing before with a computer," says Taylor. With use of this strategy, the Harvard team found the fifth-known pulsar. Within a year they found nearly half a dozen more, a sizable jump in number. And by then theorists at last figured out what a pulsar is.

It was agreed that a pulsar is a neutron star, an object first imagined in the early 1930s. Soviet theorist Lev Landau had initially suggested that the compressed cores of massive stars might harbor "neutronic" matter. The neutron, a major constituent of the atom along with the proton and electron, was a hot item at the time, having been recently discovered by experimentalists. Caltech astronomers Walter Baade and Fritz Zwicky picked up on the idea and proposed that under the most extreme conditions—during the explosion of a star to be exact—ordinary stars would transform themselves into naked spheres of neutrons. But their proposal was considered wildly speculative, and only a handful of physicists even bothered to ponder the construction of such a star. They would be so tiny that no one figured such miniscule stars would ever be detectable anyway. Jocelyn Bell proved them wrong.

A neutron star squeezes the mass of our Sun into a space only 10 or so miles in diameter. This occurs when the core of a particularly massive star runs out of fuel. No longer able to withstand the force of its own gravitation, the core collapses. The core had encompassed a volume about the size of the Earth, but in less than a second it becomes one giant atomic nucleus the size of Manhattan. All the positively charged protons and negatively charged electrons are com-

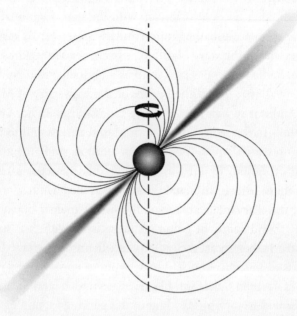

As a neutron star rapidly spins, radiation beams outward from its magnetic poles. Sweeping around like a lighthouse beacon, these beams are detected on Earth as clocklike pulses (hence the term *pulsar*).

pressed to form a solid ball of neutron particles. But like a compressed coil, the newly squeezed core rebounds a bit, generating a powerful shock wave that eventually blows off the star's outer envelope as a brilliant supernova.

The neutron star that remains behind spins very fast. Like an ice skater bringing in her arms—compressing herself to whirl ever faster—a collapsing stellar core spins faster and faster during its compression as it conserves angular momentum. Its magnetic field grows very intense as well, to a trillion gauss or so. (The Earth's magnetic field, by comparison, is a feeble half a gauss, a hundred times weaker than the strength of a toy magnet.) Such a rapidly spinning and highly magnetized body becomes no less than an electrical generator. As a result, narrow and intense beams of electromagnetic waves are emitted from the neutron star's north and south magnetic poles. As on Earth, these poles don't necessarily line up with the star's rotational axes. So as the star spins around, the beams regularly sweep across earthbound telescopes, much the way a lighthouse beam regularly skims across a coastline. Radio telescopes pick it up as a periodic radio pulse. In the fall of 1969 Taylor joined the faculty of the University of Massachusetts at Amherst to help establish the Five College Radio Astronomy Observatory in the woods of western Massachusetts and to continue his pulsar research, which was an opportunity to study the final stages of stellar evolution. By then astronomers had detected a few dozen pulsars, but to truly understand them they needed a much larger sample. Researchers wanted to know how they were distributed through the Milky Way. Could all pulsars be associated with past supernova explosions? Just studying the radio pulses themselves seemed futile. "I have a friend in India who said that trying to understand a pulsar by looking at its radio pulses was like trying to understand the innards of a complicated factory by standing in the parking lot and listening to the squeaks of the machinery. In some cases, only a tenth of a percent of a pulsar's energy comes out in radio," notes Taylor. To get answers, Taylor wanted to double or triple the number of known pulsars by extending his computerized method for finding these radio beacons. He seemed destined for such a task since childhood.

Taylor grew up in the 1940s on a farm along the New Jersey shore of the Delaware River, just north of Philadelphia. Perhaps it was the farm machinery, but he and his older brother Hal became avid mechanics. They fiddled with all kinds of motors, both gasoline-driven and electrical. They even erected ham radio antennas on the roof of their family's three-story Victorian farmhouse. Taylor's interest in electronics continued through his undergraduate days at Haverford College, where he built a radio telescope for his senior thesis. He was able to detect the Sun and about five radio galaxies, some of the most distant objects then known in the universe. In Massachusetts, though, pulsars became Taylor's passion. He recognized that, with a large sample of pulsars on hand, astronomers could begin to use them as tools for probing interstellar space, seeing how the radio signals slowed, scattered, and polarized as they traversed the diffuse gases between the stars. Perhaps there were even new species of pulsars yet to be revealed. In his funding proposal to the National Science Foundation (NSF), Taylor did note—almost as an afterthought—that "even *one* example of a pulsar in a binary system [a pair of stars in orbit around one another] . . . could yield the pulsar mass, an extremely important number." It was a minor wish; he figured the odds would be against him. All the pulsars detected so far were solitary creatures. Since neutron stars are the remnant cores of exploded stars, it seemed reasonable to assume that the explosion would have disrupted the orbit of any companion star. Convinced of the merits of a large computerized pulsar search, though, the NSF allotted $20,000 for Taylor's project, a sizable sum in its day.

Needing assistance, Taylor sought out Russell Hulse, a graduate student then looking for a thesis project. Taylor offered him the perfect dissertation topic, a survey that would combine all three of Hulse's top interests: radio astronomy, physics, and computer science. Hulse readily accepted, entitling his project, "A High Sensitivity Search for New Pulsars." Like Taylor, Hulse had been a tinkerer since his youth. When he was nine years old, he helped his father build a summer vacation home in upstate New York, putting in walls, rafters, and siding. "I was always building things," he says. "Fortunately, I came through the experience with all my fingers intact." An eclectic child, he first went through his chemistry and biology phases, dissecting frogs and mixing chemicals. By the age of 13, electronics had captured his fancy; at that time he was also admitted to the legendary Bronx High School of Science, notable for its Nobel Prize–winning graduates. Sparked by a library book on amateur radio astronomy, he built a radio telescope out of old television parts and army surplus in the backyard of his family's summer home. "Electronics was a lot more accessible than it is now," he recalls. "If you opened up a radio or TV set, there were all these parts: resistors, capacitors, tubes, wires, and coils. And you could take these electronic parts and build an antenna that had the potential to detect radio waves from the Milky Way." It was magical for him, the idea of detecting signals out of the ether. "I don't have cable TV even today," he says. "I still get my signals the old-fashioned way, out of the air with an authentic antenna." His hand-made telescope consisted of two flat sheets, each about 4 feet by 8 feet, covered with wire mesh and meeting at right angles. Strung down the middle were a couple of dipole antennas. He tuned it to 180 mega-hertz, what would be channel 8 on a television dial. It didn't work, but he was never bored by the attempt. Through these experiences he developed a freewheeling, "I'll-do-it-myself" attitude that served him well in his schooling. He later taught himself programming on an early IBM computer at his undergraduate institution, Cooper Union, a college in lower Manhattan. One of his first programs was an orbital simulation.

Hulse chose the University of Massachusetts at Amherst for grad-uate work so he could combine his interest in electronics with astron-omy. "Radio astronomy was still new, still rough-and-tumble," he says. And UMass, as it's familiarly called, was building a new radio tele-scope, what turned out to be an array of four 120-foot-wide radio dishes. He arrived on campus in 1970. By the time he was ready to tackle his thesis three years later, pulsars were still being found using a hodgepodge of techniques. His and Taylor's plan was to conduct a more systematic search taking advantage of the latest technology—called a "minicomputer," although still as large as a couple of microwave ovens—to dig deeper into the galaxy for both weaker and faster pulsars. This required the use of the Arecibo Observatory in Puerto Rico, the largest single radio telescope in the world. More than three decades in service, this telescope is legendary for the range of its observations. It made the first accurate measurement of Mercury's rotation, discovered planetary systems outside the solar system, and listened for signs of extraterrestrial life. Nestled in a natural bowl-shaped valley in central Puerto Rico, the telescope was initially built to explore a layer of the Earth's upper atmosphere known as the iono-sphere. That required an antenna 1,000 feet in diameter, encompass-ing the area of a dozen football fields. A limestone sinkhole in a valley near the town of Arecibo provided a natural framework for such a mammoth structure. The large collecting area also made it perfect for picking up pulsar signals, which are very weak.

The minicomputer, a Modcomp II/25, was programmed to sweep across a wide range of possible pulse periods and pulse widths in assembly-line fashion as the radio telescope scanned the sky overhead in Puerto Rico. The aim was to look for a range of pulsars, ones that beeped as fast as 30 times a second or as slow as once every 3.3 seconds. They also looked over a range of dispersions, the amount of spread between a pulsar's high and low frequencies. All in all there were half a million possible combinations. "At each point in the sky scanned by the telescope," notes Hulse, "the search algorithm examined these 500,000 combinations of dispersion, period, and pulse width." This made the search 10 times more sensitive than previous surveys.

The computer was housed in two crude wooden boxes, a combination packing crate and equipment cabinet that Hulse had built out of plywood. It had 32,000 bytes of core memory, a goodly amount for its day but thousands of times less than today's desktop computers, which now routinely incorporate tens of millions of bytes. A teletype was used for input and output, while a reel-to-reel tape drive stored the data. To get the maximum processing speed possible, Hulse programmed the computer in assembly code—the machine's internal digital language—using 4,000 punch cards, an experience he does not long to repeat.

Hulse carted the minicomputer to Puerto Rico at a fortunate time. The telescope was then undergoing a major upgrade. Many observations were impossible at this time, but pulsar searching could still be carried on. This afforded him more time for his searches than he would have received normally, working in and around the construction and other people's observations. In fact, he stayed at Arecibo for some 14 months—from December 1973 to January 1975—with only the occasional trip back to Massachusetts for a break.

The huge Arecibo dish does not move. It just serenely watches as the heavens continually turn above it. To look for a potential pulsar signal as the telescope carried out its passive sweep, Hulse examined a particular spot on the sky for 136.5 seconds. Then he would begin to examine the next spot over. Hulse's prime-time viewing each day was when the plane of the Milky Way, toward the inner parts of our galaxy, passed overhead for some three hours.

Just before that critical time, Hulse conducted the same routine. First he ran a tape, loading his program into the computer's memory. "As a classic computer hacker—and I'm using hacker in the positive sense—I had to make this program run fast enough so that all the data it collected over that three-hour observation window could be processed within 24 hours, before the next observation came up," he says. "I spoke fluent hexadecimal." During the observation itself, the computer would carry out the dispersion analysis and write its streams of digitized data onto a big magnetic tape. Over the remaining hours of the day, some 21 hours, the computer would review the resulting data and look for telltale signs of a pulsar. If the computer found a suspect, it would awaken the teletype and have it type out a cryptic line of information, which Hulse could easily translate. "You would know immediately," says Hulse. "The teletype started going *chunk-chunk-chunk-chunk*. It had a long line to print out if it found something. If there were interference, it was a nuisance, then you'd get all sorts of junk. Paper would start overflowing." False leads could arrive from nearby thunderstorms, which were common during the summer in Puerto Rico. There was one pesky candidate signal that turned out to be emanating from an aircraft warning light on one of the telescope's support towers. And then there were the days that the U.S. Navy held exercises off the coast. "I just sat in the control room," recalls Hulse, "watching signals from the naval radars . . . jump around on the observatory spectrum analyzer." But Hulse learned quickly to distinguish a false signal from a real one just by looking at the teletype print-out.

By the end of his 14-month stay, Hulse had cornered 40 new pulsars, all located in the roughly 140 square degrees of the Milky Way observable with the big Arecibo dish. With each new find he drew a hash mark on the side of his trusty Modcomp II/25. All in all he quintupled the number of known pulsars in that particular sector of the sky. That alone made a nice thesis for Hulse. But "it was of course eclipsed by the discovery of what was to become by far the most remarkable of these 40 new pulsars, PSR 1913 +16," he points out.

PSR is astronomical shorthand for pulsar, while 1913 is the pulsar's right ascension in celestial coordinates. It stands for 19 hours and 13 minutes. Astronomers have divided the sky into 24-hour segments, akin to longitude but in this case the time it takes the Sun to make one complete circuit. The 16 is the pulsar's declination or latitude on the celestial sphere. That placed the pulsar midway between the Aquila and Sagittarius constellations, close to the galactic plane that passes overhead at Arecibo. Hulse had started perusing this sector in the summer of 1974.

Things were pretty routine at that point. Hulse had already found about 28 pulsars and by then even had pretyped forms to fill in the pertinent information on his finds. July 2 started out as just another day until one particular signal squeaked by the threshold Hulse had set—just barely. With a bit of interference he would never have seen it. It was the teletype, automatically reporting any interesting finds, that first informed Hulse. It was an unusual candidate, as its signal was particularly fast, with a period of about 58.98 milliseconds (17 "beeps" a second). "It would be the second-fastest pulsar known at that time, which made it exciting," says Hulse. (The faster pulsar was then the famous 33-millisecond pulsar [30 beeps a second] situated in the Crab Nebula, the remains of a supernova that was seen to explode in 1054.) Being such a weak signal, though, Hulse was still skeptical. "I put it on my suspect list," he says. "After the list got long enough, I'd devote a whole session reobserving them." Weeks later he was able to confirm the source. He proceeded to write down the signal's characteristics, tacking on a flourish at the end: "Fantastic!" he wrote on the bottom of his discovery sheet. His contact with Taylor was irregular, since phone service on the island often didn't work and e-mail was decades away. Through regular mail Hulse let his advisor know that he may have discovered a fast pulsar.

Hulse got back to PSR 1913 +16 and his other suspects once again on August 25. This time it was the opportunity to measure their periods—the rates of their radio pulsing—more accurately. For the most part this was a standard and easy procedure: he just measured the candidate once and then measured it again about an hour or so later, to gather extra data for a more accurate measurement. But for PSR 1913

+16 the period actually changed over that hour. The two measured periods differed by 27 microseconds (0.000027 second). "An enormous amount," says Hulse, at least for a pulsar. "My reaction . . . was not 'Eureka—it's a discovery' but instead a rather annoyed 'Nuts—what's wrong now?' " Figuring it was an instrument error, he simply went back and measured it again another day. He kept marking down a new period on his discovery sheet, one after another. After the fourth one he just scratched them all out in frustration. Obtaining accurate pulse periods was *not* a requirement for his thesis, but his compulsive nature took over. Perhaps his equipment wasn't sampling the pulsar fast enough to obtain an accurate fix on its period, thought Hulse. He then spent a full week writing a special computer program for the Arecibo mainframe to handle a faster data stream. He dropped all his other investigations and for two days solely observed this persnickety pulsar. But the problem only got worse. "Instead of a few data points that didn't make sense, I now had lots of data points that didn't make sense," says Hulse. Yet he did notice some regularities. He saw that the pulsing rate had decreased; the next day it decreased yet again. "My thesis wasn't going to fail if I didn't do this measurement, but beyond some point it became a sheer challenge," says Hulse. "I couldn't live with myself until I understood what was happening."

His thinking shifted at this point. He became convinced that the pulsar's period was actually changing, that it wasn't just an instrument error. He spent hours visualizing a spinning pulsar, trying to imagine how it might slow down. Finally, the image of a *binary* pulsar came to mind. Perhaps his undergraduate experience simulating stellar orbits paid off at last. At that point Hulse didn't know that Taylor had mentioned the possibility of finding such a system in the NSF proposal. In such a binary the pulsar would be orbiting another star. And that's why the pulsar's period varied. The pulsar period would regularly change—rise and fall, rise and fall—due to the orbital motion. When the pulsar moves toward the Earth, its pulses are piled closer together and its frequency appears to rise slightly; when moving away from us, the pulses get stretched and the frequency decreases. Optical astronomers have been acquainted with this effect for decades when observing the visible light of binary star systems. An audio version happens right here on Earth as well—the familiar rise and fall in the pitch of a train whistle, as the train first races toward us and then away.

In his gut Hulse knew that he was right, but he had to see the "turn-around," the moment when the pulsar started approaching the Earth in its orbit. If the pulsar were indeed a binary, its frequency should at some point start to increase. Finally, on September 16, he saw it. His notebook records the proof. He had been processing the data in five-minute intervals, and every time the computer arrived at a period for that five-minute span, he marked it down on his graph paper. "I clearly remember chasing the period. Every one of those dots was a separate little triumph. The real exultation was seeing it hit the bottom and then turn around. There wasn't any doubt that it was a binary system. I drove back home that night, down the winding roads from the observatory, thinking 'Wow, I don't believe this is happening.'" It was also a relief. The chase had been stressful and had delayed work on his thesis.

Fairly soon, Hulse could tell that the pulsar was orbiting another object roughly once every eight hours. He quickly mailed Taylor a letter—a remarkably grumpy letter, actually—moaning about the extra work the pulsar had created for him. Today, Hulse says the isolation and lack of sleep had probably gotten to him. But even with the letter on its way, he decided the news couldn't wait. With telephone connections so difficult from Arecibo, Hulse used the observatory's short-wave radio link to Cornell University. Cornell, in turn, patched the call via a phone line to Amherst. Taylor, immediately recognizing the import of Hulse's find, got someone to take over his classes and flew down to Puerto Rico within a couple of days with better pulsar-timing equipment.

Taylor and Hulse soon confirmed that the two objects were orbiting one another every 7 hours and 45 minutes. That meant they were moving at a rather speedy clip, about 200 miles per second, a thousandth the speed of light. While one was surely a neutron star, because of the pulsing, the other was likely a neutron star as well, since it was not big enough to eclipse the pulsar. (Pulsing is not detected in this second star because its beam is most likely not aimed at the Earth.)

The size of the binary's orbit is not much bigger than the radius of the Sun, a relatively slim 435,000 miles. A light beam could cross it in two seconds. With such intriguing properties, Taylor and Hulse immediately recognized that they had been handed on a silver platter the perfect relativistic test bed. Hulse remembers going to the library at Arecibo fairly soon after his find and consulting a copy of Misner, Thorne, and Wheeler's *Gravitation*. In their paper announcing the discovery, they wrote that the "binary configuration provides a nearly ideal relativity laboratory including an accurate clock in a high-speed, eccentric orbit and a strong gravitational field." Over just a few months they could actually detect the orbit of the binary system precessing, slowly dragging around. "That's the analog of the change in Mercury's orbit, but in this system it's much larger," says Taylor.

Up until the discovery of the binary pulsar, tests of general relativity were primarily carried out within our solar system. But with PSR 1913 +16, the entire galaxy opened up to experimental testing of the rules of space-time. When Einstein first derived the formula indicating that two objects orbiting one another would release gravity waves, he also recognized that the two objects would be drawn closer and closer together, due to the loss of energy that the waves carry off into space. PSR 1913 +16 was the perfect candidate to test this out. Here were two test masses—so compact and so dense—continually moving around one another. It was the ideal setup for detecting gravity waves (at least indirectly). Imagine a twirler's baton spinning in a pool of water. The motion would create a set of spiraling waves that move outward. Similarly, the motion of these two neutron stars should emit waves of gravitational energy that spread outward from the system. With energy leaving the binary system, the two neutron stars would then move closer together. At the same time their orbital period would get shorter. "As they depart for outer space, the gravitational waves push back on the [stars] in much the same way as a bullet kicks back on the gun that fires it," Thorne once explained. "The waves' push drives the [stars] closer together and up to higher speeds; that is, it makes them slowly spiral inward toward each other." But seeing such an effect required great patience. It could not be observed immediately but only over years.

While Hulse went on to other endeavors, Taylor and several col-

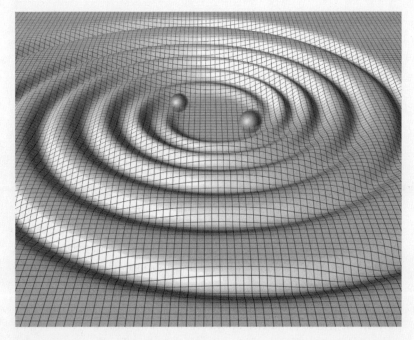

As two neutron stars orbit each other, they generate gravitational waves in space-
time that spread outward like ripples in a pond.

leagues, particularly Joel Weisberg, now at Carleton College in Minnesota, continued to travel to Arecibo to monitor the evolution of PSR
1913 +16. Year by year they would spend two weeks—sometimes
more—measuring the system as precisely as they could. Their major
goal was to pinpoint the pulsar's timing. The tick of this pulsar clock is
very regular, a sharp pulse every 0.059 second. Its blips are so regular
and stable that its accuracy can rival the most accurate atomic clock
on Earth. But to detect any changes in the binary's orbital motions, via
the pulsar's precise tick, required extraordinary measurements. The
system is located some 16,000 light-years away, so its signal is very
weak. Taylor's group had to build a special receiver that could better
analyze the signal. It took four years of monitoring before they could
finally detect a very slight change in the orbit of the two neutron stars.
The answer arrived after analyzing some 5 million pulses. The orbit

was definitely shrinking. The two stars were revolving around each other a little faster. That meant the binary system was losing energy and the neutron stars were drawing closer together. Moreover, the energy loss was exactly what was expected if the system were losing energy in the form of gravity waves alone. It was a tough problem in relativity. Taylor first used an approximation that he found as a homework problem in a classic textbook on general relativity. But even before handling the calculation, he and his group had to make a number of corrections to the data. They had to correct for the motion of the Earth in the solar system as well as the perturbations introduced by the other planets. Variations in the Earth's rotation affect the signal's timing. There is also a slight delay of the signal due to the interstellar medium. "We measure and remove, measure and remove," says Taylor. They even had to adjust for the motion of the solar system around the center of the galaxy.

The gravity wave news was first released at the Ninth Texas Symposium on Relativistic Astrophysics, held in Munich, Germany, in December 1978. (The conference series originated in Texas, hence the name.) It was the highlight of the meeting. A report came out two months later in the journal *Nature*. But, initially, there were doubts. Some wondered whether there was a third object in the system, which would upset the calculations. Or maybe dust and gas surrounded the pair, which could also explain the energy losses. But additional measurements over the years—with better and better receivers—only improved the accuracy. Taylor's graph, plotting the ever-decreasing orbital period, is a showpiece of science. The measured points lie smack dab on the path laid down by general relativity. The measured energy loss due to gravitational radiation agrees with theory to within a third of a percent. Such accuracy has been described as "a textbook example of science at its best." Each year the binary's orbital period decreases by about 75 millionths of a second. During each spin around each other, in the continuing *pas de deux*, the two neutron stars in PSR 1913 +16 draw closer by a millimeter. Over a year that adds up to a yard. The two stars will collide in about 240 million years. So clean and precise is this system that Taylor once remarked it was "as if we had designed the system ourselves and put it out there just to do this measurement."

In 1981 Taylor moved to Princeton University, but he continued to glean information on PSR 1913 +16 from the simple ticks of its clock. After a couple of decades of measurement, some of the relativistic changes are fairly dramatic. The binary's orbital precession, for example, is quite vigorous. Taylor and his group now measure the change as 4.2 degrees per year; that's 35,000 times larger than the annual change in Mercury's orbit. The reason is clear: two neutron stars, so close together, affect the warping of space-time far more than our less dense Sun. Tightly bound together, they pack quite a wallop. Since its discovery, the binary pulsar has shifted its orbit a full quarter turn. Using this information, along with other orbital parameters, Taylor and his colleagues have been able to peg the masses of the two neutron stars to four decimal places. One is 1.4411 solar masses; the other is 1.3873. That's quite an accomplishment from a distance of 16,000 light-years, given that each neutron star is a superdense nugget only 10 miles wide. "One has to marvel at how much is learned from so sparse a signal," says Rainer Weiss.

The Hulse-Taylor binary, as it is now called, is no longer the sole member of its species. More than 50 binary pulsars are now known, with most of the pulsars paired up with white dwarf stars rather than neutron stars. These particular pulsars spin much faster and so were not seen right away with older equipment. Stealing material from the white dwarf, the pulsar whirls itself up to faster and faster speeds, like a twirling ice skater on overdrive, hundreds of times each second. Hulse left it to others to discover these new binaries. He actually left the field of radio astronomy just a few years after his momentous find. He was not looking forward to the wandering life of a postdoc and the scarcity of secure academic positions. Wanting to be near his girl-friend, he took a job at the Princeton Plasma Physics Laboratory in 1977, where he continues to work as a principal research scientist on computer modeling. Hulse hasn't changed much since his graduate school days. He still has his dark hair and beard. And he remains an ebullient and engaging talker, like a young kid explaining his favorite hobby. Taylor, meanwhile, continues his pulsar work, in and around his new duties as dean of the faculty at Princeton. His tasteful office in historic Nassau Hall is decorated with an antique grandfather clock, which he dutifully winds every morning, even though it is far less accurate than his beeping neutron stars. The team's old minicomputer is gone, long since cannibalized for parts, but Hulse does retain his original printouts, on newspaper-like green paper. "It's such hacker stuff. I read it now in a daze," says Hulse with a chuckle. "I enjoyed doing it *once*."

Hulse had sighted the first pulsar of his extended search on December 8, 1973. Exactly 20 years later to the day, he was at a podium in Sweden delivering a lecture on the work he did to garner his Ph.D. He and Taylor had just received the 1993 Nobel Prize for Physics for their masterpiece of measurement, one of the few times that prize over its century-long history has been awarded to astronomers. In the lecture Hulse described his work as "a story of intense preparation, long hours, serendipity, and a certain level of compulsive behavior that tries to make sense out of everything that one observes." He didn't ignore a troublesome observation. He tackled it with fervor, finding for Taylor and the astrophysics community the perfect laboratory for relativistic physics.

There had been controversy when Jocelyn Bell Burnell was denied a share of the 1974 Nobel Prize in Physics for her role in the discovery of the pulsar. The coveted award went to her advisor, Hewish, instead. This antistudent bias changed with the discovery of the binary pulsar. "It was very much a joint effort," says Taylor, a man well known in the astronomical community for his generosity and gentlemanly spirit. "Yes, one of us was a student, but there was no question that Hulse's work was an essential part of the operation." A longtime friend of Bell Burnell, Taylor invited Jocelyn to accompany him and his wife to Sweden for the award ceremonies.

Bars and Measures

Throughout the 1960s a certain question was regularly heard drifting through the hallways at general relativity conferences: "Has Joe Weber seen anything yet?"

Years before Joseph Taylor started obtaining indirect evidence for gravity waves, Joseph Weber was resolutely trying to catch one directly. It was a utopian crusade, and he knew it. At the time he started on his venture at the University of Maryland, his fellow physicists expected it would require a full century of experimental work to attain such a goal. Even Weber admitted that "the probability of success under these circumstances had to be regarded as very small." But they admired his moxie.

Before Weber no one had even contemplated pursuing such an experiment. And there was good reason: "It had to scare any sane person," says Peter Saulson, for a gravity wave is such an incredibly tiny effect in our local surroundings. It was once calculated that if the great ocean liner *Titanic* were spinning once a second, it would generate less than a million billion billionth of a watt of gravity power. An atom bomb just 10 yards from a detector would generate a gravity wave signal more than a trillion trillion times weaker than the one you could detect from a distant supernova exploding in our galaxy. It shows how difficult a task it is to create gravity waves that are detectable enough to run an experiment. Only events on a cosmic scale can provide the sources.

By convention the strength of a gravity wave is usually stated in terms of its "strain," a term borrowed from engineering. It's the fractional change in length—the magnitude of stretch—the wave would impart either in the distance between two masses or in a block of material. Two black holes merging in the center of our galaxy would certainly emit a sizable series of gravity waves. In fact, they would be deadly if you happened to be there. The waves would alternately stretch and squeeze any object nearby by as much as the object's size. A 6-foot man would be stretched to 12 feet and within a millisecond squeezed to 3, before being stretched out once again. Ignoring all the other forces at work when black holes collide, any planets and moons in the vicinity could be torn asunder under the stress of gravity wave power alone. It's a frightening prospect. But, fortunately, by the time those waves reached us here on Earth after traveling some tens of thousands of light-years, their strain (in the metric units that are the *lingua franca* of science) would be a paltry 10^{-18} meter per meter. In other words, each meter along a rod would have its length changed by 10^{-18} (a millionth trillionth) of a meter, a span a thousand times smaller than the width of a proton.* The cosmic tsunami is reduced to a quantum quiver. It's a cumulative effect. The longer the measured span, the greater the overall effect. It builds up over distance. How

---

*For those unfamiliar with scientific notation, 10^{18} means 1,000,000,000,000,000,000, a 1 followed by 18 zeros, an enormous number. Conversely, 10^{-18}, read as "10 to the minus 18," means 1/1,000,000,000,000,000,000, an extremely tiny number. 10^{-19} is a number ten times smaller than that, 10^{-20} is a hundred times smaller than 10^{-18}, and so on.

Isaac Newton never imagined the existence of gravity waves, which alternately stretch and compress space-time as they pass by.

would such a strain, for example, affect the 93-million-mile distance between the Earth and the Sun? With a strain on the order of 10^{-18} the gravity waves would expand and contract that space-time by a span about equal to the width of a germ.

Detecting such miniscule changes seemed an impossible task, but Weber fearlessly thought otherwise. He was partly sparked by his work at Maryland, teaching electrical engineering. "It seemed to me that if you could build an electromagnetic antenna to receive electromagnetic waves, you might be able to build a gravitational wave antenna to receive gravitational waves," he said in recalling his thinking at the time. "I didn't know where they might come from. I just thought I'd start looking." In a way he wanted to be the Heinrich Hertz of gravity. Inspired by Maxwell's equations of electromagnetism, Hertz had confirmed Maxwell's prediction that electromagnetic waves existed. Weber was similarly determined to catch one of Einstein's predicted ripples in space-time. Astronomy was undergoing such a revolution after World War II, with the emergence of new disciplines such as radio and x-ray astronomy, that it was getting easier to contemplate finding the impossible. In the violent universe then being unveiled, nature might indeed be generating gravity waves capable of detection. Moreover, being the one to clinch one of the last predictions of general relativity was a powerful motivation.

Weber started seriously pursuing relativity research during the

1955–56 academic year while on sabbatical at the University of Leiden in the Netherlands, where John Wheeler was spending some time, and at the Institute for Advanced Study in Princeton, New Jersey, Einstein's old haunt. Both J. Robert Oppenheimer, then the institute's director, and Wheeler encouraged Weber's new interest. Later, his hopes would be nurtured by institute physicist Freeman Dyson, who calculated the gravitational waves emitted by the collapsing stellar core at the heart of a supernova. Dyson's results suggested at the time that the gravity wave signal would be far stronger than previously expected. Weber published his daring scheme for detecting gravitational radiation in 1960 in *Physical Review*. In that and subsequent papers he outlined a clever technological trick for trapping a gravity wave. He surmised that a burst of gravitational energy moving through a solid cylinder would alternately squeeze and expand it ever so slightly, like an accordion in motion. The change would be incredibly small—as mentioned earlier, far less than the width of a nuclear particle. But then, long after the wave passed through, the bar would continue to "ring." This phenomenon is similar to the vibrations that can be produced in a tuning fork when it is struck by sound waves. Similarly, a gravity wave that is "tuned" to the bar's natural acoustic frequency sets off a resonance, much like a gong continuing to ring after being struck. In both cases the dimensions and material of the bar or the gong determine which frequency of wave will trigger the ringing. Weber reasoned that he could position electronic devices on the sides of the cylinder and convert the extremely tiny gravity wave–induced movements into electrical signals that would then be recorded and scrutinized. This was not the first time Weber ventured into virgin scientific territory. Charting new waters was in his bones.

Weber was born in Paterson, New Jersey, in 1919 to Jewish immigrants and named Jonas Weber. His father was Lithuanian, his mother Latvian. The family name was originally Gerber, but that changed when his father, eager to move to the United States, took the visa of another man who had decided at the last minute to stay in Lithuania. With his family's prime language being Yiddish, Jonas mistakenly turned into Joseph when his mother first registered him for school. Like many of his generation, Weber became fascinated by radios as an adolescent. He obtained his ham radio operator license at the age of 11. Working as a golf caddie during the Depression for a dollar a day, he saved enough money for a book on electronics and started repairing radios after school. He was the youngest of four children, which he said was a blessing. While his older siblings were put to work early to help support the family, he was left alone to focus on his schoolwork. He decided to enter the U.S. Naval Academy to break away from his immigrant roots and applied to a New Jersey senator, who relied on tests to make his appointment. Weber excelled and got the slot. Soon after graduating in 1940 with a degree in engineering, he was thrust into World War II. Assigned to the aircraft carrier *Lexington* as its radar officer, Weber shipped out of Pearl Harbor on December 5, 1941, two days before the Japanese attack. The following year he survived the carrier's sinking in the Battle of the Coral Sea. He finished out the war as commander of a submarine chaser in the Mediterranean. He also married his high school sweetheart, Anita Straus, with whom he had four sons.

After the war Weber extended his mastery of the new science of radar technology as head of the electronic countermeasures section of the Department of the Navy's Bureau of Ships. So great was his expertise that the University of Maryland hired him in 1948 at the age of 29 as a professor of electrical engineering, with the stipulation that he also pursue his Ph.D. He did this at nearby Catholic University, where he obtained his degree in microwave spectroscopy in 1951. While completing his doctorate he worked out the concept of what later came to be known as a maser, the trailblazing forerunner of the laser. Instead of emitting visible light, a maser generates a beam of pure microwave energy—electromagnetic waves intermediate in length between infrared and radar. Weber says he was inspired by a course in atomic physics. He was the first to publicly mention the maser principle at a meeting in Ottawa in 1952, the Electron Tube Research Conference, a yearly gathering devoted to the cutting edge of electronics. Weber never built his proposed device, though. For one, his calculations suggested that its performance would be minimal. Moreover, he had no research funds to construct a working model. Nobel Prizes went to those who did—the U.S. physicist Charles Townes and two

Russians, Nikolai Basov and Aleksandr Prokhorov—based on alternate mechanisms: "I was only a student in a way," said Weber. "I didn't know how the world worked."

Hearing visiting scholars lecture on general relativity at the university, Weber decided to use his 1955 sabbatical to study the subject in more depth. He did this, he says, because he "had wide interests and no money for quantum electronics." During this time he worked with Wheeler on the theory of gravitational radiation. It was a time when the debate was still raging on whether gravity waves truly existed or were just an artifact of the mathematics. "My philosophy," said Weber, "was to act like Galileo: build something, make it work, and see if you find anything." He spent nearly two years thinking of scheme after scheme for catching a gravity wave. He filled four 300-page notebooks with possible detector designs. His 1960 *Physical Review* paper presented his most promising method for a gravity wave receiver: a gravity wave would hit an object—specifically, a piezoelectric crystal—whose vibrations would be converted into electrical signals and recorded. Piezoelectric crystals have the interesting property of generating a voltage when squeezed. Quartz is such a crystal. Pierre Curie first noticed this effect (and named it) in 1880. Finding it too costly to obtain a sizable bar of piezoelectric material, Weber and his lab associates came to realize they could bond the crystals to a much larger mass, a solid aluminum bar, which was cheap, easy to work with, and vibrated well. For a while Weber also wondered whether the entire Earth could be used as a detector. The Earth, if "plucked," has various modes of oscillation, starting at one cycle every 54 minutes. Using a gravimeter, an instrument that measures the motion of the Earth's surface, Weber hoped to "tune into" the Earth's natural frequencies and see if it had been excited by a passing gravity wave. But background noises such as earthquakes, the motion of the oceans, and atmospheric events simply overwhelmed any potential signal. (Weber did persuade NASA to have Apollo 17 astronauts place a gravimeter on the Moon, where some thought there was a better chance of detecting a wave because of the Moon's lack of weather, quakes, or oceanic disturbances.) But in the end Weber concluded that his best bet was building a detector in his laboratory.

Robert L. Forward, an employee of Hughes Aircraft Research

Laboratories based in California, had arrived on the Maryland campus in 1958 to begin work toward a doctoral degree. Originally intending to do research on masers, Forward heard about Weber's pioneering venture and signed on. He obtained his Ph.D. by constructing the first gravitational wave antenna, along with David Zipoy. According to Forward, choosing its size was straightforward: "The reason for its particular length is that I said, 'Well, if I am going to have to manhandle that monster, let's make it this big.' I spread out my hands to what I could grasp, and we decided to make it 5 feet." Its width was 2 feet. The bar's total weight came to about 2,600 pounds. It was a fortunate choice of size. As soon as neutron stars and black holes moved from the realm of science fiction to real science over the 1960s, crude calculations (which largely still hold up) suggested that gravity waves emanating from these new celestial creatures would likely have frequencies of a few thousand hertz (a few thousand waves passing by each second), frequencies that could be picked up by solid cylinders of aluminum several feet long. (Before that the Maryland team had expected its future bars would have to be hundreds of yards long to record the longer waves emanating from ordinary binary star systems.) To isolate it from seismic disturbances, the bar was suspended on a fine steel wire in a vacuum tank, which rested on acoustic filters that shielded it from environmental disturbances. Surrounding the "waist" of the cylinder, like some bejeweled belt, were the piezoelectric crystals. According to Weber's strategy, once the bar was deformed by a wave—a gravitational burst—it would continue to oscillate and convert those vibrations into electrical signals. That's why these systems are also known as resonant bar antennas. The bars resonate like a bell in response to the passing gravitational wave.

Forward went back to Hughes in 1962, upon completing the bar's construction. Afterward, the bar operated on the Maryland campus from 1963 to 1966. Pulses were recorded on a chart recorder and scanned by eyeball. (Later, computers would be brought in for better hands-off analysis.) But interpretation was difficult with one bar alone. The very motions of the atoms in the bar (on the quantum level, atoms are continually shaking) can swamp any gravity-induced signal. Weber believed he could get around this problem by operating two bars simultaneously. He reasoned the chances were low that the atoms would rattle around in each bar in the exact same way at the exact same time. Tremors occurring in both at once would then likely be due to an outside disturbance.

Additional bars were built and placed in a garagelike laboratory, what Weber called his gravitational wave observatory, about a mile away. Small numbers of coincident pulses were observed. These results were reported in 1968, but interpretation was difficult. When a car accidentally ran into his outpost, generating a huge pulse, Weber realized that his detectors needed to be much farther apart, so local disturbances could be ruled out as a possible source of interference as well. "Suppose you see a big pulse on one detector," noted Weber. "You can't be sure whether that pulse was due to a garbage truck colliding with the building, a lightning strike, or student unrest." (During the turbulent 1960s, student protesters painted obscene graffiti on his observatory.)

Weber's realization led to the construction of two identical bars. Each was 5 feet long and 2 feet wide and weighed about 3,100 pounds. One was stationed on the Maryland campus, the other at the Argonne National Laboratory near Chicago, some 700 miles westward. The two detectors were linked by telephone line. If one of the detectors surpassed a certain threshold—what was determined to be the thermal background noise—the system was programmed to emit a pulse. If the other detector crossed that threshold at the same time, within 0.44 second, a coincidence marker was triggered. Most of the time the needle traces on their ever-moving chart paper displayed what was expected: random wiggles. But for a brief instant in December 1968, the needles on both detectors jumped at the same time. Over the next 81 days, Weber's team observed 17 significant events shared by the two separated detectors. The researchers went through an exhaustive process to make sure the signal was real. They inserted time delays into their electronics to rule out random coincidences. If the signals were truly random, then inserting a time delay in one bar's circuits wouldn't have affected the coincidences at all. But the count fell, suggesting it was more than chance at work. They made sure it wasn't some kind of electromagnetic disturbance, such as a solar flare or lightning. They monitored cosmic-ray showers and seismic disturbances at each site. In their estimation, nothing could explain the coincidences except the momentary passing of a gravitational wave. The size of the cylinders determined the frequency. It was 1,660 hertz (cycles per second), in the range of frequencies expected to emanate from an exploding star.

It was 10 years of work altogether, first arriving at a strategy and then setting up his instrumentation. It was difficult for Weber to keep the discovery under his hat. He made his move at the Relativity Conference in the Midwest, a meeting held in Cincinnati in June 1969 and attended by America's top relativists. Kip Thorne was there, giving a report on the possible gravitational radiation generated by newly formed neutron stars. "And then Weber got up and announced that he had seen gravitational waves. It was quite a shocker to people," recalls the Caltech physicist. Thorne had known Weber for years and was highly interested in the techniques he was pioneering. "So I took it very seriously, as did nearly everyone else," he says. Conferees greeted the announcement with applause and tributes. Two weeks later Weber's official report was published in *Physical Review Letters*, the journal of choice for quick dissemination of a major discovery throughout the physics community. For a time Weber was headline news. His picture was hard to miss, with his piercing eyes, resolute mouth, and a precision crewcut that had his hair standing at attention. The popular press wondered aloud in its stories whether Weber's discovery was the most important event in physics over the last half century. Weber's laboratory certainly became a magnet and an inspiration for physicists over the following months. Signing his visitor logbook in 1970 were William Fairbank of Stanford University and Ronald Drever from the University of Glasgow, who would soon establish their own gravity wave detection programs.

Weber was particularly interested in applying his find to astronomy. His announced detection came only two years after pulsars had been discovered, and he already had his eye on that phenomenon as a possible source for his gravity waves. By noting the period when his events were most prolific, Weber concluded that the gravity waves were arriving from the center of the Milky Way galaxy. His proclamation was eerily reminiscent of Karl Jansky's announcement in 1932 that radio waves were emanating from the galactic center, an event that marks the birth of radio astronomy; at the time no one had ever expected such radio energies to be emanating from the galactic core. According to the Weber team, the position of its two detectors offered a crude method for tracing the source of the alleged signal. At each site the long axis of each cylinder faced east-west. Situated this way, the strongest signals would occur when the source was straight overhead (or straight below the Earth, which poses no obstacle to a wave). Thus, the Sun could not be the source, they concluded; the signal didn't noticeably increase at noon or midnight when the Sun was directly overhead or below. According to their initial statistics, the gravity wave signals did peak when the center of our galaxy, located in the direction of the constellation Sagittarius, was in those prime positions. No one knew what events might be causing the "blips" on Weber's moving graph, but there were guesses: supernovas going off in the galaxy, colliding neutron stars, or matter falling into black holes, a term just coming into use.

Initial media coverage of Weber's announcement, even in the specialized physics press, was highly enthusiastic. It seemed that everyone was caught up in this new endeavor. It added a bit of élan to the field of general relativity. Here was a novel technique to plumb the depths of the universe in a whole new way. Suddenly, any conference on relativity (normally a quiet affair) became a hot ticket; it was practically standing room only as people gathered to hear the latest news, exchange ideas, and compare notes. The response was not unlike the rush in 1989 to confirm the purported discovery of cold fusion. Within a year of Weber's discovery, at least 10 groups were either proposing or already mounting similar searches. Work began in the Soviet Union, Scotland, Italy, Germany, Japan, and England. In the United States, gravitational wave detector groups were eventually assembled at Bell Laboratories in New Jersey, IBM in New York, the University of

Rochester, Louisiana State University, and Stanford University in California. Due to their fixed size, Weber's detectors were limited to recording one frequency, 1,660 hertz. The newcomers' goal was to make detectors both more sensitive and tuned to additional frequencies, in order to extend the science. At the time there were also a number of competing theories of relativity avidly being discussed, and some of them, like the Brans-Dicke theory, predicted different effects when a gravity wave passed through a bar. So some hoped to use their new gravitational wave detectors to test whether Einstein was right or wrong.

One of the first to put a detector together and check Weber's claim was Vladimir Braginsky, Dicke's counterpart in the Soviet Union. Stimulated by Weber's prediscovery publications on gravity wave instrumentation, Braginsky began to write papers on possible sources and methods of detection as well. Early on he recognized that thermal noise—the incessant jostling of a bar's very atoms—would produce major interference. By 1968 he was testing other types of bar materials, such as sapphire crystal, to see if this noise could be reduced. Given this head start, he was able to construct a detector fairly quickly after Weber's pivotal announcement. Braginsky's group at Moscow State University later imagined building a series of bars of differing sizes—a sort of xylophone—to register different frequencies of waves at the same time. More than that, the Soviets began thinking of rocketing detectors into space. One scheme involved taking a detector shaped like a dumbbell far above Earth and rotating it. Theory suggested that a passing gravity wave would alter the rotation.

Elsewhere other imaginative schemes were being devised. In Colorado researchers led by Judah Levine of the Joint Institute for Laboratory Astrophysics (JILA) ran tests in an abandoned mine, the Poor Man Relief mine located several miles west of Boulder in Four-Mile Canyon. There they reflected a laser light between two mirrors, mounted on piers 30 yards apart. It was a crude setup. Supposedly, a standing wave could be set up between the mirrors, which a disturbance would alter. The changes would be noted by comparing the wave with a laser set to a constant frequency. R. Tucker Stebbins worked on the laser experiment while a graduate student at Colorado. "It used the Earth as the resonant bar, with a laser interferometer to detect the resonance," explains Stebbins. The instrument had already been set in place earlier to measure the speed of light more precisely, which required a quiet spot. The gravity wave experiment ran a year or two but detected nothing, except for some earthquakes and underground nuclear tests. The seismic interference was far too overwhelming.

Besides the solid cylinders of aluminum, now called Weber bars, designers came up with new configurations, such as hollow squares, hoops, and U shapes. Others, such as Fairbank, world renowned for his work in low-temperature physics, began thinking about cooling the bar. The detector would be placed in a dewar, in effect a huge Thermos bottle. By lowering the bar's temperature to diminish atomic jitter, its sensitivity could increase thousands of times over Weber's room-temperature system. Meanwhile, new researchers in the field began to switch to a new type of sensor to increase their sensitivity. A small diaphragm was placed at the end of the bar. Physics decrees that any energy oscillating in the big mass will eventually transfer to the tiny mass on the end. But with that energy entering a smaller mass, the motion amplifies, making it easier to measure. While these new strategies were being developed, Weber worked on improving his own detectors.

Theorists were not immune to the excitement surrounding this newfound field. They were immediately drawn to figuring out what Weber was seeing, including British theorist Stephen Hawking. One of his earliest scientific papers analyzes the type of signal a gravitational wave detector could register. He and his coauthor Gary Gibbons argued that the sort of bursts recorded by Weber could be coming from stars undergoing gravitational collapse to become neutron stars. But in working out the numbers, they cast one of the first dark clouds on Weber's work. They had to question the myriad events being reported. For the energies being detected, the newborn neutron stars had to be located within 300 light-years of Earth. But by then the Maryland group was saying it was seeing an event each day; there simply weren't that many neutron stars nearby. What if the feeble signals were actually emanating from the Milky Way's center, some 30,000 light-years distant, as Weber was claiming? Then the energies had to grow tremendously. The energies had to be high enough to maintain any detectable strength by the time the waves reached Earth in the far suburbs of the galaxy. It was necessary for each pulse, radiating outward from the galactic core, to involve the conversion of roughly a Sun's worth of mass into pure gravitational energy. But with such events occurring daily that meant the galaxy would be losing mass at a tremendous rate, far too rapid to keep our galaxy intact from its birth more than 10 billion years ago to the present day. The galaxy would be imploding at its center, long on its way to gravitational annihilation. If that was truly the source of Weber's signals, the galaxy should have been gobbled up by now. With such results theorists started to become highly skeptical of Weber's assertions. "Either Joe Weber was wrong," commented one theorist, "or the whole universe is cockeyed."

By 1972 William Press and Kip Thorne wrote a review article of the field's accomplishments that posed the possibility that Weber might be wrong. But they did not yet reject Weber outright. They did offer some alternate possibilities: for one, the energy of the alleged gravity wave source might be less than it appeared. Maybe an unknown source was nearer or the radiation somehow "beamed" in only one direction. Or perhaps the gravitational radiation in the universe was far stronger than previously suspected. Could it be originating from outside the galaxy? "If these excitations are caused by gravitational radiation," wrote Press and Thorne, "then the characteristics of *each* burst are about what one expects from a 'strong' supernova or stellar collapse somewhere in our Galaxy; but the number of bursts observed is at least 1,000 times greater than current astrophysical ideas predict! Weber's observations lead one to consider the possibility that gravitational-wave astronomy will yield not just new data on known astrophysical phenomena (binary stars, pulsars, supernovae) but also entirely new phenomena (colliding black holes, cosmological gravitational waves, ???)." Their question marks left the door open to the unexpected.

By then the experimentalists were beginning to experience grave

doubts about Weber's discovery as well. Braginsky registered no signals and simply shut his detector off. He felt his energies would be better spent developing more sensitive detectors. David Douglass at
Rochester and J. Anthony Tyson at Bell Laboratories also came up
empty-handed. Tyson had been fascinated by the possibility of detecting gravity waves for nearly a decade, ever since he read a small mono-
graph by Weber entitled "General Relativity and Gravitational Waves"
as a graduate student at the University of Chicago. Upon completion
of a postdoctoral project in 1969, he was hired by Bell Labs to conduct
research in low-temperature physics, his specialty, but Weber's historic
announcement that year was too tempting to ignore. To corroborate
Weber's detection, Tyson secretly set up two small bar detectors—each
about 3 feet long and 1 foot wide—in his laboratory, which was large
enough for added equipment to get lost in the clutter. Though his bars
were smaller than Weber's, he did use extremely low-noise amplifiers.
Tyson ran his detectors covertly for about a year. "And I didn't see a
thing," he says. At last getting his employer's blessing to continue the
investigation, Tyson built another bar that was appreciably larger than
Weber's original detectors. It was roughly 12 feet long and 2 feet wide
and weighed almost 4 tons. According to Tyson, it was capable of
observing gravity wave pulses that were far weaker than the events first
recorded by Weber. But after steadily operating for a month in the
summer of 1972, it registered no special signal. Any surge that did
appear was not beyond that expected from the random motions of the
bar's atoms. Furthermore, Tyson arranged to observe the galactic center with one of the optical telescopes at the Cerro Tololo Observatory
in Chile at the very same time he was running his test. Nothing out of
the ordinary was observed; no visual bursts matched up to the events
Weber reported he was seeing over that same period. Based on the signal Weber was allegedly receiving, says Tyson today, "there should have
been enough energy in other forms, such as electromagnetic waves, to
knock your socks off. All you should have needed were binoculars."

Reporting his findings at the 1972 Texas symposium held in New
York City that year, Tyson got into a heated argument with Weber.
Tyson had been supplied with four months of data taken earlier by

Weber's group. The Bell Lab team tested for correlations with sunspots, temperature and barometric pressure differences near Weber's Maryland laboratory, as well as earth strains midway between Argonne and Maryland because of lunar and solar tide effects. In the process Tyson and his colleagues discovered that with high probability the detected signals over that period aligned with changes in the Earth's magnetic field at the equator, an anomaly known as the disturbed storm-time factor. It is thought to be related to the ring current in the ionosphere around the equator. This was not proof that the Earth's magnetic field was the source of Weber's signals, but it opened up the question of whether some of Weber's events were of terrestrial origin.

Responding to Tyson, Weber stressed that his group did run magnetic tests on his bars, with field strengths far larger than the Earth's, and that his bars did not respond. He also countered that detecting a signal requires at least two detectors as "coincidences are the result of externally produced signals, small compared with the noise of each detector." You must compare to see anything, he said. At that same meeting Weber was upping the ante. He reported that his group was now seeing two or three coincidences per day.

Tyson conceded that he didn't as yet have a second detector for comparisons but stressed that his one detector was receiving nothing; there were no glitches of any kind above the normal noise levels. He was particularly distressed by Weber's lack of calibration. Weber had not yet tested his bar with a known source of energy to determine the exact strain it could see. Tyson himself used an electrostatic calibration. He applied a set force to the bar and noted the response. Amid this growing criticism, Weber continually offered one major rebuttal: all the other groups were simply not building the same instrumentation. He firmly believed the sensors had to be placed around the waist of the bar, not on its side. "The scheme that works is not very difficult to set up," he declared. "[Until] you have reproduced the experiment that is known to work, I don't think I believe anything." He also contended that his competitors had inadequate temperature controls in their labs, burying any signals in excess noise.

Observers sympathetic to Weber's efforts sounded a similar refrain. Perhaps only Weber's bar, they noted, was actually "tuned" to the events erupting daily. Cardiff University sociologist Harry Collins, who has been conducting a sociological study of the field's development for nearly three decades, even came across some overreaching supporters who briefly wondered whether psychokinesis was at work, with Weber as a focus.

The Maryland physicist did have more expertise. By then Weber had spent a dozen years designing, building, and testing his equipment before claiming his first signal. His competitors had spent, in some cases, less than a year. A colleague of Weber's told Collins that "one of the things that Weber gives his system, that none of the others do, is dedication—personal dedication—as an electrical engineer, which most of the other guys are not."

But eventually other bars identical to Weber's were constructed, and the news was not good for the founding father of the field. A collaboration in Europe—one group in Frascati, Italy, the other in Munich—built detectors that closely matched Weber's original design, including having the sensors wrapped around the bar's waist. One of their tests ran for 150 days, intermittently from July 1973 to May 1974. Expecting to see at least one pulse a day, like Weber, they ended up seeing nothing at all. The barrage of these negative reports eventually reached a critical mass, erupting in a confrontation that has become legendary in the history of the field.

Richard Garwin, a maverick physicist at IBM known for his scientific crusades and a gravitational wave novice, decided to build an antenna to settle the issue once and for all. Garwin, who in his early twenties had helped design the first hydrogen bomb, was very suspicious of Weber's statements and wary of his statistics. With a colleague at IBM, James Levine, he built a 260-pound detector within six months. Running a month in 1973, the small aluminum bar picked up one pulse, likely a noise. Garwin later learned from David Douglass that at least some of Weber's daily signals may have been the result of a computer mistake. Douglass had noticed a programming error that would register a coincidence between Weber's two antennas, when none actually occurred. Nearly all the "real" coincidences reported by Weber over one five-day period could be traced to this error. He was seeing a signal in what most likely was pure noise. Weber was also claiming to see coincidences between his detectors and an independent detector run by Douglass in Rochester. But that couldn't have been possible. The two labs, it was discovered, used different time standards. One lab used Eastern Standard Time, the other Greenwich Mean Time. The data that Weber had been comparing (and which appeared to show coincidences) were in reality four hours apart. For Garwin and others this seemed to prove that Weber was selectively biasing his data to suit his claims.

In a surprise attack, Garwin verbally confronted Weber with this information at the Fifth Cambridge Conference on Relativity held at MIT in June 1974. An acrimonious exchange ensued at the front of the lecture hall. As the two approached each other with clenched fists, moderator Philip Morrison, disabled by childhood polio, raised his cane to keep them apart until the tension subsided. Their battle continued, though, in an exchange of terse letters in *Physics Today*, the magazine of record for the American Institute of Physics. Garwin contended that the very way in which Weber defined a coincidence introduced errors. Analyzed in a different manner, the purported signal went away. Garwin and his associates carried out a computer simulation to demonstrate this effect. Though their data were random, they could coax what looked like a signal out of sheer noise.

For many, this confrontation reminded them of an earlier contretemps in 1970, when Weber first claimed his signals were emanating from the galactic center. At a lecture, he announced that the peak reception occurred every 24 hours, at a time when the Milky Way's center was overhead. Then someone in the audience pointed out that Weber's reception should also be good when the galactic center was directly below his antennas (the Earth being no barrier to a gravity wave). It wasn't long before Weber was reporting that his peaks were indeed arriving every 12 hours. (To boost the event rate for a better statistical analysis, Weber contends his group had been "folding" the second 12 hours of data over the first 12 hours since those scans above and below the Earth match each other. This, he says, resulted in his initial misstatement.)

Though not backing down from his conclusions, Weber did initiate a number of checks and balances at his laboratory both before and after this confrontation with his critics. First of all, he removed himself from direct participation in the data reduction to eliminate any personal bias. He put it into the hands of graduate students, postdocs, and associates. Any data recorded on pen-and-ink charts or signals picked out by eye were now cross-checked with automatic computer systems. Artificial pulses were inserted and identified by programmers, unaware of the timing. Telephone circuits connecting Maryland and Argonne were checked for spurious noises. From Weber's point of view, he answered all criticisms and corrected all possible sources of error, but the damage was done. From that point on, his announcements and papers were viewed with growing distrust by others in the field. They were leery of his scientific methods and frustrated that his publications at times were hazy on details.

Just two weeks after the infamous Cambridge meeting, the Seventh International Conference on General Relativity and Gravitation was held in Tel Aviv. The final session gathered four key participants in the bar experiments: Joe Weber, Tony Tyson, Ron Drever from Glasgow, and Peter Kafka from Munich. Kafka reported on the negative Munich-Frascati results, while Tyson talked about his latest effort. In collaboration with colleagues from the University of Rochester, he had constructed an even larger bar and began operating it in 1973. This bar would eventually run, off and on, for some eight years. Partway through that period, its twin was built by Douglass at Rochester, and the two detectors ran in coincidence for some 440 days from 1979 to 1981. Again, not one cosmic shiver was registered. Tyson jokingly called his last gravity wave antenna "the most expensive thermometer in the world."

During the Tel Aviv panel discussion, Drever reported on his ongoing work in Glasgow. His device was fairly different from others then in operation; he set up two separate masses and had them linked by piezoelectric transducers. Each mass weighed about 600 pounds.

He and his colleagues conducted a seven-month run and saw no sign of a signal. "My aim," he told the conferees, "was *not* to try to find out if Weber's work was right or wrong but to find out more about gravitational waves." Meanwhile, Weber used his time to defend his methods, answering his critics over computer programming, his choice of algorithms in analyzing data, and the calibration of his detectors.

At the end of the session, Drever pondered the meaning of the differing results. "You have heard about Joseph Weber's experiments getting positive results. You have heard about three other experiments getting negative results and there are others too getting negative results, and what does all this mean? . . . [Is there] any way to fit all of these apparently discordant results together?" There were potential loopholes. Drever's experiment was not sensitive to long pulses. Perhaps that's why he was missing a signal. Kafka and Tyson were sensitive to certain waveforms, the ones most expected by theorists. But maybe nature wasn't acting according to the script. On the other hand, Drever's design would have caught unusual waveforms, since it could capture a broader range of frequencies. "I think that when you put all these different experiments together, because they are different, most loopholes are closed," he concluded.

For some the Tel Aviv conference marked the end of their involvement in gravitational wave physics. Their interest swiftly waned when Weber's findings were put into doubt. But the field was hardly reaching an endpoint. To the contrary, the atmosphere in Tel Aviv was one of great excitement and eager expectations. Other participants felt they had only scratched the surface of the field's potential. Many were excitedly talking about the possibility of increasing the sensitivity of the bars to detect supernovas out to the Virgo cluster of galaxies, some 50 million light-years away. So instead of waiting for a Milky Way supernova to go off every 30 years or so, they would enlarge their territory and possibly catch several events a year. To do that, though, sensitivities had to increase a million to 10 billion times over 1974 standards. But such a sizable leap was hardly deterring the newcomers to this new brand of astronomy. Now that they had dipped their toes into the water, they were ready to plunge in headfirst.

Nearly everyone was coming to agree that Weber was mistaken. "We are not certain, but it is probable," noted Drever in his concluding remarks in Tel Aviv. But he pointed out that, even with null results, the field had broadened. People were beginning to bubble with ideas on what else could be done to improve their chances of seeing a bona fide wave. Some were choosing the low-temperature route. Others thought of using crystals with large resonances, enabling the detectors to ring for longer times. "Another technique which is coming into view now," noted Drever, "is the quite different possibility of having separate masses which are a long distance apart. . . . One may monitor the separations using laser techniques." But their desire was more than to just push the technological envelope. Science was always on their minds. "From a confirmed initial discovery that can be reproduced readily, I think the thing could rapidly spread to where we would have a real astronomy and we would be producing maps of the sky of gravitational wave sources," said Drever. "Every time we have looked into the sky with a new kind of detector, a new black box, we have found something which we did not expect," added Tyson.

In later years Weber would occasionally have his hopes built up. A week's worth of data from the University of Rome, taken with a supercooled bar in July 1978, were compared to Maryland data taken at the same time with one of the room-temperature bars. In a 1982 paper in *Physical Review D*, a claim was made that correlations were seen between the two detectors. Weber immediately announced that this was additional confirmation that he and others were seeing gravity waves. The Rome researchers, however, preferred to describe it as a background. Their conclusion in the paper was cautious: "The observation of a small background of coincidental excitations tells us nothing concerning the origin. Detection is statistical. There is no way of separating the coincidences which are due to chance and those due to external excitations. And we cannot be sure what fraction of the external excitations are of terrestrial or nongravitational origin."

Despite their doubts over Weber's methods, even his harshest critics recognized the tremendous engine that the Maryland physicist had set into motion. "It is clear," said Tyson in 1972, "that if it were not for

Weber's work, we would not be as near as we are today to the possible detection of gravitational radiation . . . with the use of supersensitive, low-noise antennas." Weber had created a momentum that could not be stopped.

Within 10 years of the Tel Aviv conference, scientists had developed the second generation of bar detectors, each cooled with streams of liquid helium to a chilly –456° Fahrenheit, near absolute zero. This cuts down on the thermal noise inside a bar, which creates motions hundreds of times larger than a gravity wave displacement itself. For a while one of the most ambitious of these projects was located at Stanford University, under the guidance of both William Fairbank and Peter Michelson. Their 5-ton aluminum bar was situated in a hulking metal tank within a vast room that was once an end station for the original linear accelerator laboratory at Stanford. When conditions were good, the supercooled bar could at times detect a strain of around 10^{-18}. That meant it could register a shiver that changed the bar's dimensions by one part in a billion billion. This was a 10,000-fold improvement over Weber's first instruments. The use of superconducting instrumentation, as well as the cooling of the bar, contributed to this sensitive performance. Such sensitivity put supernovas exploding all over the galaxy within their reach. Unfortunately, the 1989 Loma Prieta earthquake seriously damaged the Stanford detector. Too expensive to rebuild from scratch, the entire operation closed down.

But the work continues at Louisiana State University, which had been working hand in hand with Stanford and had a detector identical to the Stanford bar, 10 feet long and 3 feet wide. In 1970 William Hamilton, a protégé of Fairbank's, had arrived at Louisiana State to supervise the construction of both bars at a nearby NASA facility just past the border in Mississippi. He is now teamed with Warren Johnson in running the gravity research effort at LSU. Around 1980 they switched to a new aluminum alloy with a far better resonance. Slimmer than their old bar at a mass of 5,000 pounds, this detector is known by the musical name of Allegro, which stands for "A Louisiana Low-temperature Experiment and Gravitational Radiation Observatory." It's an apt moniker for an instrument listening for a gravity wave's tones, which happen to fall in the audio range. The bar is tuned to 907 hertz and at its best can detect a quiver smaller than 10^{-18} meter.

In a fluorescent-lit basement laboratory at LSU, an enormous vacuum chamber sits in a corner, its pumps regularly chugging away. The lab is a prime example of organized clutter: cranes, ladders, barrels are scattered pell-mell. Inside the chamber the Allegro bar anticipates a gentle push from space-time. It has been a long wait. Banks of electronic equipment sit right by the tank, monitoring the bar's every movement. If a wave does come in, it will cause the ends of the bar to go in and out, ever so slightly. Those small movements will then be transferred to a secondary resonator attached to one end; this causes the tiny motion of the big mass to translate into a bigger motion of the small mass. Nicknamed "The Mushroom," because of its shape, this resonator amplifies the signal. Allegro first came on the air in 1991 and, except for a short period to perform an upgrade, has been on ever since—24 hours a day, 7 days a week, a significant achievement. Its operation has served as an incubator for training a whole new generation of bar experts. "We're not seeing gravity waves," says Hamilton, "but we do have some enigmatic noises." Adds Johnson, "We have several events per day, things outside normal stochastic noise, but we can usually track them to local occurrences." On one spring day some pile drivers working down the street could be detected in the continual computer record kept on the bar's output. Allegro is not alone. Similar ultracold bars have been operating around the world. Together, they form a gravity wave detection network. The frequencies to which they are tuned range from roughly 700 to 1,000 hertz. The Louisiana State detector is joined by the "Nautilus" in Frascati, Italy, near Rome; the "Auriga" at the Legnaro National Laboratories near Venice; the "Explorer" at CERN (European Center for Nuclear Research) in Switzerland; and in western Australia the "Niobe" (so named because it is made of niobium, a metal more resonant than aluminum. Indeed, the niobium can ring for several days once activated). All are capable of seeing either a supernova go off or two neutron stars merging within our galaxy, but they also have a chance of detecting other events as well, though far rarer. Out to a distance of a million light-years (halfway to our closest spiraling neighbor, the Andromeda galaxy), they might be able to observe two massive black holes colliding. While up and running, they are attempting to track down and comprehend all possible sources of interference: electromagnetic disturbances, cosmic rays, and seismic tremors. The Nautilus, cooled to within a tenth of a degree of absolute zero, has actually recorded the vibrations generated by sporadic bursts of energetic particles passing through the bar, showers created whenever cosmic rays strike Earth's atmosphere. Once such noises are understood and subtracted out, bar researchers can then focus on the unexplained events. Data comparisons are under way. Allegro and Explorer, for example, have had their data compared from a 107-day period in 1991. No coincidences were revealed, but scientists at both sites are now looking for continuous waves that might arrive from pulsars. Tucanae 47, a globular cluster with lots of pulsars in it, might be a rich source. Indeed, all the detectors are beginning to coordinate their searches and check for identical signals. Data from Explorer and Nautilus were compared over a period when there were a number of gamma-ray bursts, which are suspected to involve the collision or explosion of massive stellar objects in the far universe. As yet no unambiguous gravity wave signals have been noticed, but the search continues.

The next advance in bar design is an ambitious concept: a spherical detector. For a while several projects were being planned. In the Netherlands a consortium of Dutch universities and institutes (Amsterdam University, Eindhoven University, Leiden University, Twente University, and NIKHEF, the Nuclear Physics and High Energy Physics Institute) had organized the Grail Project. This would have been a 3-meter-wide sphere, 110 tons of copper alloy, suspended and cooled to within thousandths of a degree of absolute zero. The huge sphere was to be cast by a company experienced in manufacturing ship propellers. To minimize cosmic-ray impacts, it was to be placed underground, at least half a mile down. (Both the added size and the increased sensitivity of the sphere make cosmic-ray impacts far more troublesome.) Their biggest challenge would have been getting efficient cooling and yet still remain quiet enough to listen for a passing gravity wave. The vibration would have been detected by placing a set of sensors over the surface of the sphere, a placement that would enable it to see a signal from any cosmic direction. Lack of funding, however, canceled the project.

But Italy is carrying out a similar endeavor called the Sfera Project. Physicists at the INFN (Istituto Nazionale di Fisica Nucleare) in Frascati hope to construct a 100-ton sphere composed of copper and aluminum. They expect to build a smaller version at first, perhaps a 10-ton prototype. If this new approach is finally launched, it could reignite the field of resonant detectors. The increased sensitivity that a sphere would offer, perhaps a level 10,000 times better than current bars, has the potential to make it competitive as an observatory that could study far more than supernovas.

The bar detectors now in operation have a very good chance at seeing a supernova, if the explosion is close enough. As luck would have it, though, no second-generation detector was "on the air" on February 23, 1987, when the light from Supernova 1987A, which had exploded in the Large Magellanic Cloud right here in our own galactic neighborhood, at last arrived. And, on average, astronomers only get to see such nearby events every 30 to 50 years. Bar detectors are not yet a working science but still more like an ongoing technological project. The advanced bars were temporarily offline to make improvements. But certain room-temperature bars were working. Although Weber by then had ceased publishing his data regularly, he did occasionally surface to make a report. At the American Physical Society's spring meeting in 1987, he announced that one of his detectors registered excess noise—vibrations stronger than background—for a few hours around the time of the explosion. A gravitational wave detector operated at CERN by University of Rome physicists also reported seeing events at that time. The alleged pulses occurred over a period during which particle detectors situated in Italy's Mont Blanc, in an Ohio salt mine, and in Japan recorded an extra burst of neutrinos. The probability that such a coincidence could happen, claimed Weber and the Italian researchers, was one in 1,000 to 10,000.

Was the detection real? The rest of the community is nearly unanimous in saying *no*. Other physicists have closely examined the data and have concluded that the Maryland-Rome statistics are highly flawed. For one, the Mont Blanc neutrino detection remains an enigma. The neutrino detectors in the United States and Japan saw the supernova go off four and a half hours later than Mont Blanc. It's hard to understand how a supernova—a violent and abrupt event—could persist for several hours, as the Maryland and Rome bars seemed to suggest. The Maryland-Rome collaborators even reported, "It is possible that new physics [is] needed," to explain the origin of the correlations.

Weber lost his longtime support from the National Science Foundation shortly after the announcement of his supernova claim. One gets the impression that if Weber had not taken such a hard line over these years—refusing to at least consider the possibility that his data might be spurious—he would have remained an honored member of the gravitational wave detection community rather than moved to its sidelines. "It was one of Weber's failings, but a fatal one," says Saulson. "He would stick to his guns, despite all the mounting evidence."

By 1998, two years before his death at the age of 81, Weber's hair had thinned and whitened, but he was no less a striking presence with black glasses that framed intense blue-gray eyes. For visitors he dressed formally: gray suit, white shirt, and solid red tie. He was always an avid jogger and mountain climber, boasting that he had climbed all the 14,000-foot-high mountains in Colorado. Even in his late seventies he was still trim, having maintained his Naval Academy weight, and he continued to be vigorous and quick to defend. Knowing a reporter was about to visit, he amassed an organized defense, laying out his argument paper by paper in chronological order. Textbooks now take a uniform stand on his work; all state that gravity waves have never been confirmed. He picked up a representative book and recited aloud the offending section: "Other scientists have built more sensitive instruments and yet gravity waves have not been detected." He was clearly dismayed.

Weber long held the title of senior research scientist at both the University of Maryland and the University of California at Irvine. He liked to say he was fired when he reached retirement age at 70, but he continued to keep watch over his experiment when he could. His first wife died of a heart attack after 29 years of marriage. In 1972 Weber married astronomer Virginia Trimble and began the bicoastal arrangement, spending part of the year in Irvine, where Trimble has a post, and the remainder in Maryland. His Maryland office looked more like a storage closet, filled to the brim with 17 file cabinets, which had been pushed together at the center of the room. Boxes of papers and bookcases lined the walls. Only one tiny aisle was left open to reach his small desk, situated against the far wall by the blackboard. Anyone dropping by had to sit knee to knee with Weber. Placed prominently over the desk was a picture of Einstein, one that Weber was told was Einstein's favorite portrait. It is a serious pose of the physicist as a young man that best accentuates his compelling eyes.

Weber was obviously disappointed at his inability to catch the brass ring. First with masers and later with gravity waves, he was the man who would be king. A scientist once prone to volatile outbursts at physics meetings, Weber displayed no anger during his interview, one of his last. His voice and manner remained matter-of-fact as he discussed his current point of view. He conceded that his original estimates of signal strengths, based on classical physics, were wrong, since they implied energies being emitted in our galactic core that are hard to imagine. But that did not mean there were no gravity waves, he stressed. He had since devised a new way to think about how his bars were receiving a signal. "When our work first started, it was based solely on Einstein," said Weber. "But around 1984 I started asking how Niels Bohr would have done it." In other words, he wondered whether quantum mechanics—how his bar was receiving the signal on the atomic level—might better explain what he was seeing. Just as classical theory cannot explain how a metal can superconduct or exhibit the photoelectric effect, argued Weber, so did classical theory fail in explaining how a gravity wave interacts with the bar. He claimed that a gravity wave can couple to individual atoms and, as a result, interact far more strongly. This, he said, explained how he could see events a billion times fainter than once thought: "My theory says that a bar is made up of 10^{29} atoms, coupled by chemical forces and described by quantum mechanics. It's no longer a single big mass." The atoms, according to this scheme, all work in concert with one another to amplify the signal, making it a billion times stronger than the older theory could account for. According to Weber, the excitation bounces from end to end, with his center electronics best seeing the energy with each pass. Other bars missed this effect, he contended, because their sensors are situated on the end. Few physicists, though, agree with this hypothesis. Weber wondered whether some of the groups contesting his claims spent enough time to confirm his findings. "If your objective is to show that the effect is null," he said, "you don't have to spend five years trying to get the experiment to produce a null result. You turn it on, and it produces a null result. They didn't give it enough time, enough care."

Tyson, who today keeps his hand in gravitational physics by finding and studying gravitational lenses throughout the universe, now applauds Weber's ingenuity. "Joe came up with a fantastic idea, which, to this day, is state of the art for bar detection," he says. "You have to praise Weber for figuring out how to do that." Where they parted company was in the interpretation of the bar's response. Tyson has concluded, along with others in the field, that Weber misunderstood the exact nature of the natural noises emanating within a bar, which led him to interpret what are really false alarms—simultaneous but random vibrations in the bars—as coincident signals.

Weber's observatory was still operating in 1998. He drove to it in a 1972 powder-blue Volvo (his wife's). The lab was situated at the edge of the campus in a thick grove of trees, near the college golf course. The boxy white structure could easily have been mistaken for a small garage. Inside, a small overhead bulb cast a dim light. One detector was half hidden behind a jumble of odds and ends. It resembled a giant red oil tank. A second detector was about 20 feet away against the back wall, looking like an aluminum water heater. The aisle was too cluttered to reach it easily. The state of Maryland continued to pay for the electricity and special environmental controls to maintain a stable temperature in the room. When asked about other expenses, Weber silently answered by putting his hand into his suitcoat and pulling out his leather wallet. He did receive a NASA grant at one point, which allowed him to purchase a computer to keep track of his data. His bars were continuing to operate 24 hours a day, and a data point was registered every tenth of a second. Weber paid for the storage disks himself. Once they were turned on, the vibrant computer monitor screens added the only touch of modern technology in a room filled with aging equipment. Two jagged lines—the ongoing signals from each detector—silently scrolled from right to left. The red tank contained a cylindrical bar 2 feet wide; the gray tank at the end of the room held a fatter bar, 38 inches wide. Both had been running since 1969, three full decades. Weber admitted his bars were deteriorating. The mounts were worn, and sensitivity was down. But he was determined to keep them running somehow until the latest gravitational wave observatories were online. He was sure their data would coincide. "This observatory is important," he said, pointing to the two detectors, "because it's producing data." Judging those data, however, was always a matter of great disagreement—between Weber and nearly the entire gravity wave community that he spawned.

Inspired by a suggestion from his astronomer wife, Weber went to NASA in the mid-1990s to suggest that the signals from his ever-working bars be compared to data received from BATSE (Burst and Transient Source Experiment), an all-sky monitor mounted on NASA's Compton Gamma-Ray Observatory launched in 1991. BATSE has been observing gamma-ray bursts over the entire celestial sky. Receiving a $10,000 NASA grant, Weber used it to pay for a postdoc to carry out the statistical comparison. Of 80 gamma-ray bursts registered by BATSE between June 1991 and March 1992 the postdoc found that 20 of these events coincided within half a second to pulses from the larger of Weber's bars. Weber claimed the probability of that being coincidental was one in 600,000. Some gamma-ray bursts are thought to arise when a newly formed black hole ferociously "eats" its surrounding gas, in the process shooting out an intense pulse of energy from its poles. Other bursts may occur when binary neutron stars collide, a star explodes, or a lone pulsar is hit with cosmic debris. "After the data were written up and submitted for publication, I took the list down to NASA," said Weber. "NASA identified 11 out of the sample as due to one particular bursting pulsar near the galactic center. There's every reason to believe that source has operated sporadically over the years." When asked why other gravitational wave detectors—instruments far more sensitive than his own—were not registering the same waves, Weber simply shrugged and suggested that no one was allowed to discover a signal until LIGO was up and running.

This work on gamma-ray bursts fueled Weber's hope that he would eventually be vindicated. He admitted that it was hard to explain some of the signals he had originally been seeing. "It could have been noise," he conceded, perhaps even an atmospheric phenomenon that affected both separated sites. But for him the gamma-ray bursts offered a definitive cosmic source. "I'm sure of two things," said Weber. "Death and taxes. But the evidence that I've seen gravity waves is overwhelming." NASA did not continue its support of Weber's data comparison, but the work was reported in an Italian journal, *Il Nuovo Cimento*, noted for its liberal policy of publishing controversial results. No one listened, and no one followed up. Like the reaction to the fabled boy who cried wolf once too often, the rest of the community simply ignored Weber's last work. They didn't believe that room-temperature bars, antiques by present-day standards of technology, could have been recording anything more than local noises. Outsiders were sympathetic to his tale. Scientists in the field were less understanding. No one will take away his historic stature, though. His first bar now resides in the Smithsonian Institution in Washington, D.C.

Dissonant Chords

Rainer Weiss is a dervish, whirling about his office at the speed of sound—his sounds. He's refreshingly direct and rakishly profane. One spring weekend in the midst of LIGO's construction, Weiss is hunched over his desk, worried about a graph displayed on his computer screen. It exhibits the growing density of gases leaking into the newly installed vacuum tubes at LIGO in Louisiana. "This was actually expected," he says offhandedly. "The same thing happened at the Washington site." His high-ceilinged office was then situated on the edge of the MIT campus in a ramshackle, three-story building known solely by its numerical designation: Building 20.

If buildings can pass on good fortune, Weiss will have a favorable shot at finding a gravity wave. The wooden structure was built during World War II to carry out military research on a new technology known as radar. After the war, Building 20 witnessed the founding of the modern school of linguistics under Noam Chomsky, the erection of one of physics' earliest particle accelerators, the refinement of the Bose speaker, and Harold Edgerton's astounding stop-action photography. "The plywood palace," they called it. Weiss himself punched through its flimsy ceilings and walls to help build the most accurate atomic clocks in the world. Building 20 was supposed to have lasted just through the war. With its stained siding and peeling paint, it remained standing for five decades as field mice and squirrels occasionally ran along the pipes in the corridors. Weiss's research group was the very last to leave before the building was finally demolished. It was too environmentally hazardous to preserve.

Weiss was not eager to pack up. "My whole career was here," he says, with a sweep of his arm as he walks down the silent hallways with their well-scuffed wooden floors. All the rooms are completely stripped of fixtures and furniture. Only wires and pipes remain dangling from the walls. For the moment, Weiss's pedestrian office, with its worn-out couch and linoleum floor, stands as a lone oasis. Long tables jut out into the room, each tabletop piled high with files stuffed with papers. Surprisingly, Weiss can recite the contents of each stack by heart. He's a talker. Stories, explanations, remembrances all spill out, seemingly simultaneously. He's feisty yet self-deprecating at the same time, a man who doesn't hesitate to tell you the story of how he flunked out of MIT as an undergraduate because he paid more attention to a girl than his studies. Short with thinning gray hair and an assortment of glasses to see at different distances, Weiss is a workaholic who was supposed to cut back after a heart attack (but hasn't).

First and foremost Weiss is an experimentalist. If he hadn't become a physicist, he'd be an electrician. He plainly prefers hanging out with scientists who get their hands dirty rather than those from the chalkboard brigade. It makes his blood boil to hear of experimentalists described as foot soldiers, with the generals—the theorists— looking down from the hill. "That's what those bastards want you to think," he says, with typical bluntness. "They think they own the field. They don't. Ideas come as much from experimentalists as from theorists. They're not generals. They're jerks like the rest of us. A good book has to be written on general relativity from the experimental approach. Wait until we discover black hole events. To understand what's really happening there will take more than what the theory can now provide. We're going to turn general relativity into a science, not just a description." LIGO is a completely different approach to gravity wave detection, extending and advancing the endeavor begun by Weber. Although LIGO is very much a collaborative project, involving dozens of scientists, if forced to name the founding father of the effort, many point to Rainer Weiss.*

Weiss was born in Berlin in 1932. His father, a physician, came from a wealthy Jewish family but rebelled by becoming a Communist and marrying a Protestant, a German actress. Foreseeing the political troubles ahead for Germany, the family moved to Prague. By 1939 they were one of the last Jewish families allowed to emigrate to the United States, largely due to the father's much valued medical degree. Plopped down into the middle of Manhattan, the young Weiss began dabbling in mechanics. "I took things apart all the time," he says. "Motors, watches, radios. My room was constantly a mess, and I was always getting into trouble for it." His father, who went on to become a psychoanalyst, was a bit annoyed. As a cosmopolitan German, he was more engaged with the arts and humanities, drama and literature. Weiss's one concession to his father's tastes is an interest in classical music, which he came to love as an adult and expertly plays on the piano.

By high school Weiss began fixing radios for friends and acquaintances, an enterprise that eventually grew into a business. As soon as the war ended, surplus equipment started being sold on the streets of New York. "Equipment was flooding the cities," says Weiss. "You could go downtown and buy the damnedest stuff. If you knew even a little bit, you could get the most modern technology—transformers and radar sets—for a pittance." He would either skip school or go down on Saturday and pick up vacuum tubes, capacitors, any electronic component then imaginable. When a huge fire destroyed the Paramount Theater in Brooklyn when he was 16, Weiss salvaged 10 loudspeakers, then the state of the art. He refurbished and sold them. In fact, he

---

*You might conclude that fate judges it that way: Highway 63, which goes right by the LIGO observatory in Louisiana, has long been known locally as Weiss Road.

became so successful at making and selling complete audio systems, a novel endeavor at the time, that he almost didn't go to college. His business was highly profitable. But he also had an intellectual quest: he wanted to figure out how to solve the problem of noise, to take the hiss out of a recording as the phonograph needle rubs against the shellac of the record. He specifically chose MIT so he could study to become an audio engineer and learn enough to find a solution. What he didn't count on in 1950 was MIT's strictly ordered universe, far different from his rough-and-tumble neighborhood in New York where he fixed gang members' radios to keep from being harassed. All the buildings were known by numbers. All the courses were listed by numbers. "Everything was numbers," he says. "It was completely wacky. I asked myself, 'Am I going to survive in this place?'" He almost didn't.

Bored by his engineering courses, Weiss went into physics, but his record was dismal. Hopelessly in love with a Northwestern University coed, a musician and folk dancer, he barely attended classes his junior year, spending most of his time in Illinois. When she dumped him, Weiss returned to MIT. He passed his exams, but the department flunked him out anyway for nonattendance. Despondent that he might get drafted into the Korean War, Weiss started walking around campus to clear his head. Passing by Building 20, he looked through one of its windows and saw two guys screaming at each other. "One guy was down on the floor looking at a big brass tube. The other guy was perched near the ceiling, and he was adjusting something up there," recalls Weiss. "They were trying to find a resonance in a beam of atoms, but it was hopeless." The two were working in the laboratory of Jerrold Zacharias, the scientist who constructed the first robust atomic clock for commercial use. Listening to the argument, Weiss decided the combative pair needed someone who knew electronics, his specialty. Hired on as a technician, he made himself indispensable. He found a home in Zacharias's laboratory (which both startled and chagrined his physics professors). That is where Weiss learned to design experiments, to build, and to manage. He got technical training in welding and soldering. With all the lab's high-tech equipment, he felt like a kid in a candy shop. "It was a style of physics that engendered you to *do* things," recalls Weiss with obvious fondness.

Eventually, Zacharias bent the rules to get Weiss back into school. As a graduate student, Weiss worked with Zacharias in making more and more accurate atomic timepieces. He specifically worked on a new concept, the atomic fountain, an experiment conducted right in Building 20. The idea was to send a beam of atoms upward. Like balls thrown into the air, the atoms would eventually stop and return to Earth. Once the atoms were slowed down at the turnaround, it would be easier to measure their vibrations, the very crux of an atomic clock. The setup first ranged over one floor. Weiss sent 100 million atoms upward; none were recorded as ever coming down. Weiss soon punched through the ceiling to send the atoms two floors upward, then three. He kept going higher and higher in an attempt to see at least some of the atoms, the least energetic ones, finally stopping and falling back.

Weiss worked on this device for three years, only to discover that all the atoms were moving far more energetically than anyone had anticipated. The atoms were shooting right out of the building entirely. Today, four decades later, a successful version of the atomic fountain has been built using supercooled atoms. It's not three stories tall but rather one inch high. But there was one favorable outcome from Weiss's failed experiment: he caught the gravity bug from Zacharias, who long imagined his atomic clocks being used to test general relativity. Zacharias was hoping to eventually place one of his clocks on a tall mountain in Switzerland and another in a valley some 7,000 feet below in order to measure the gravitational redshift. It would have been an early version of Vessot's experiment, only this time carried out on the Earth's surface rather than in space. Weiss started learning general relativity in anticipation. The experiment was never done, but by then Weiss was hooked on gravity and in 1962 traveled to Princeton, the hub of experimental relativity, for a postdoc with Robert Dicke.

Weiss worked with Dicke on building gravimeters to measure the Earth's unique resonances, its "ringing" when excited. His detector went off the rails during the devastating Alaska earthquake of 1963. After two years, though, Weiss was eager to return to MIT. He loved its legacy of experimental freedom. "You could think of something and then do the experiment within a couple of days," he recalls. "You could do that because there was 'junk' all around and people who knew how to use that junk. That's why MIT was such a pleasure back then." As a newly appointed MIT assistant professor, Weiss decided to measure whether the gravitational constant was changing over time, a project inspired by Dicke's alternate theory of gravity. This spurred him to work on lasers, a necessary piece of equipment for his measurement. At the same time he also began looking into the "tired photon" theory, a hypothesis (now discredited) that cosmic photons lower their frequency by losing energy as they travel through space. This introduced Weiss to the technique of interferometry, the method of choice for such a test.

In the midst of this ongoing research, the educational officer of his department asked him to teach the course in general relativity. " 'After all,' he said, 'you ought to know it,' " recalls Weiss. "I couldn't admit that I didn't know it. I was just one exercise ahead of my students." He had learned relativity as an experimentalist, not as a theorist, and so taught the course from that perspective. To understand the concept of gravity waves, for instance, he came up with a homework problem. He asked his class to envision three masses suspended above the ground, their orientation forming the shape of an L. One mass would be in the corner of the L, the others at each end. The assignment called for his students to calculate how the distance between the masses would change as a gravity wave passed by. Weiss understood that as a gravity wave moves through space, it doesn't simply compress everything in its path and then, as it passes, expand it again. Rather, it has a multiple effect. It does two things concurrently in different directions. The wave compresses space in one direction—say north-south—while simultaneously expanding it in the perpendicular direction—east-west. Though a gravity wave is a distortion of space-time, it conserves volume. The phenomenon is somewhat akin to the squeezing of a balloon: press in on a balloon's sides and the rubber will immediately bulge out from its top and bottom, in a direction perpendicular to the squeeze. Gravity waves impose a similar effect on space-time. If a gravity wave were to come straight down on an L-shaped setup, passing through the Earth, the masses in one arm would squeeze closer together while the masses in the other arm would move farther apart. This distortion can be visualized by looking at the weave in a piece of cloth as you pull it along one dimension. The squares of the weave distort in just this way. A millisecond later, as the gravity wave continues onward, this effect would reverse, with the compressed arm expanding and the expanded arm contracting. In the course of working out this homework assignment, Weiss came to see that it could be a doable experiment, especially given his recent work with lasers and interferometry. He figured that laser beams, bouncing back and forth between the masses at each end of the L, could monitor those expand/contract flutters. Here was a completely different way to detect gravity waves.* What Weiss was imagining was a modified form of the instrument Michelson used in his attempt to detect the ether.

A continuous stream of light from a laser would enter the corner of the system and be split into two beams, each directed down an arm of the L. Mirrors affixed to the center and end masses would then bounce the beams back and forth. (Later, the mirrors themselves became the test masses.) The beams are eventually recombined, at which time they optically "interfere" (hence the term laser interferometry). The beams initially could be set so that their wave patterns are "out of step." In this way, when added together, the two beams would cancel each other out. The peak of the light wave in one beam is added to the trough of the light wave in the other beam, resulting in a null signal—darkness. But if a gravity wave caused the masses to move, the two laser beams would travel slightly different distances. In that case, when the length of one of the arms changes the tiniest bit, the beams will be more in step and produce some light when combined. The

---

*Many frequently ask whether such a measurement makes sense. If a gravity wave alternately stretches and compresses everything in its path, wouldn't it also stretch and compress the laser beam as well, making it impossible to measure a change at all? The answer lies in remembering that the speed of light never varies in a vacuum. The length changes in the arms are real and are revealed by the fact that the light takes longer to travel in one arm while simultaneously taking a shorter time to travel through the other arm. It is better to think of the light being used as a clock, not a ruler, in measuring the change. For a more complete explanation I recommend reading Peter Saulson's article "If Light Waves Are Stretched by Gravitational Waves, How Can We Use Light as a Ruler to Detect Gravitational Waves?" in the *American Journal of Physics*, volume 65, June 1997.

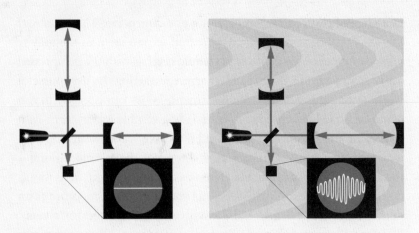

Laser light enters the system and is reflected in each arm to keep tabs on the length. When no gravity wave is present (left figure), the two arms of a laser interferometer are equal in length. When a gravity wave passes through (right figure), one arm contracts while the other expands, resulting in a signal. The movement here is highly exaggerated. The length changes will be smaller than the width of an atomic particle.

properties of the gravitational wave are hidden within those light changes. Weiss came to see that the light should bounce back and forth many times before being recombined and compared, for the repeated ricochet would increase the total distance traveled and so magnify the difference enough for sensors to detect. The sensitivity would increase even more as the masses were moved farther and farther apart.

Weiss was actually rediscovering an idea that had been in the air but not widely discussed in the gravitational wave community. Several years before Weber even announced his purported signals, two researchers in the Soviet Union suggested using an interferometer instead of a bar to detect gravity waves. But the article, published in a Soviet journal in 1963 by Mikhail E. Gertsenshtein and V. I. Pustovoit, received no notice whatsoever. Others in the field were completely unaware of its existence until Kip Thorne uncovered it many years later. "Gertsenshtein was the ultimate Caspar Milquetoast. He was unbelievably shy and mild. He had a number of seminal ideas that were totally ignored by the world," says Thorne. Weber independently thought of the idea as well. Although he did not publish it, he did discuss it with his protégé, Robert Forward, who drew a rough sketch of Weber's scheme in his lab notebook. Weiss later came up with the concept on his own when he was inspired by his classroom exercise. The idea became more and more popular over the years because, as the field evolved, it became quite apparent that bars had a number of limitations. The supercooling, for one, can be tricky. If something goes wrong, it can take several months to warm up the detector, fix it, and cool it back down again. And a bar's size limits the range of signals it can pick up. Fixed in length, a bar antenna can be tuned to only one frequency. If it were an optical telescope, it would be seeing only one color and no other, which limits its view. For many of these reasons, researchers began to focus their attention on the more versatile laser interferometer, which has the flexibility to carry out long-term astronomical investigations. It registers not just one frequency but a whole range.

A NASA astronaut served as the catalyst in getting Weiss's idea out of the classroom and onto the drawing board. In 1967 Philip Chapman had received his Ph.D. at MIT in instrumentation, with a focus on general relativity. Going on to become a scientist-astronaut, he was on the lookout for gravity experiments to conduct in space and consulted Weiss, who had been on his thesis committee. "We were going to the Moon, NASA had plenty of money, and anything seemed possible," recalls Chapman of that era. Weiss told Chapman about his idea to use laser interferometry to search for gravity waves. Chapman himself had been thinking of approaches other than bars. Enthused by the prospect and looking for further collaborators in industry, he then talked to Forward, who was continuing his bar work at Hughes. (He had three small bars operating up and down the coast of California, one of them in his bedroom closet.) "Phil Chapman put me on to Weiss's interferometer idea," says Forward, resurrecting the notion he first heard from Weber years earlier. The possibility of a future NASA project got both Forward and Weiss working on the idea independently.

Weiss's role was highly significant in this venture, for he presciently envisioned three decades ago nearly all of the crucial pieces of the laser interferometer observatories presently coming online. Carrying out an extensive design study through 1971 and 1972, the first serious examination of the technique, he identified the fundamental sources of noise with which researchers are now struggling. Moreover, he completely outlined the approaches needed to control such noises. "I was trying to be like Dicke, who would initiate an experiment by first sitting down and thinking it through completely," says Weiss. Weiss's thorough analysis, published as an MIT Quarterly Progress Report, is now viewed as a landmark paper, which is still consulted today. It is ironic that his career began with a determination to get rid of the noises in a hi-fi system, only to transfer that interest to reducing the noises that could mask a gravity wave, whose wavelength happens to be in the audio range.

Meanwhile, Forward began to construct a small prototype. He and his colleagues, Gaylord Moss and Larry Miller, spent three years building and enhancing their system. The interferometer was located in a basement room at the Hughes Research Labs in Malibu. Two aluminum pipes, usually used for irrigation in farming, were set at a right angle and served as the laser beam tubes. Each arm was 2 meters long* and aligned to be most sensitive to radiation emanating from the galactic center (where Weber's signals were then supposedly originating). The masses, set in the corners of the L, each weighed a couple of pounds. The entire system was mounted on a granite slab, set on air mounts for cushioning (an earlier setup rested on an inner tube). That was their greatest problem: isolating the instrument from various acoustic and ground noises. It was designed to receive gravity waves over a wide range of frequencies, from 1,000 to 20,000 hertz. As Forward pointed out in a journal article, they hoped that widening the bandwidth would give "significant insight into the nature of the source."

This tabletop system operated for 150 hours during the nights and

*Since interferometers are designed using the metric system, those units will be used in describing each instrument's length.

weekends from October 4 to December 3, 1972, a time when the center of the galaxy was in prime alignment. The night shift had to be used because of the high levels of noise in the lab during regular business hours. Data collection was fairly tedious and required extraordinary effort. The researchers had to sit nearly motionless for hours at a time as they monitored the interferometer, so as not to introduce any extraneous noises. The output was recorded on a stereo tape recorder, and they listened in with earphones. "Gaylord Moss and I took turns spending the night 'observing,'" notes Forward. "I found it helped to keep my eyes closed and think as if I were part of the apparatus." One channel recorded the photodetector's output, while the other channel was used to monitor environmental interferences, such as noises in the laser beam, motions of the floor, any clatter in the lab, or audible sounds from the power lines. And in the background, like a rhythmic metronome, was the incessant *tic, tic, tic* of the National Bureau of Standards time signal, broadcast by radio station WWV. This was to make sure any potential event could be timed to the nearest thousandth of a second.

At times various tones and clicks could be heard rising above a continual white noise hiss. Most of these sounds could be traced to either noises from the laser or thermal and mechanical contractions in the equipment. But occasionally, about once every 10 minutes, there would be a distinct "chirp" from the interferometer that could not be traced to any internal noise or outside disturbance. None of these signals, though, were picked up by bar detectors operating at the same time. "In view of Weber claiming to have seen gravity wave events," says Forward, "I believed it was worthwhile to operate the interferometer as an antenna for a few months, just to see if there was anything there. I did, and whatever leftover noises I heard on the interferometer were *not* Weber events."

Chances were extremely slim that such a small prototype, the first of its kind, would have detected a cosmic signal anyhow. To improve its response, Forward had plans to take his interferometer to a remote site and extend its arms to much longer lengths, possibly a kilometer or more. An optical telescope gains more resolution and sensitivity by going to bigger and bigger mirrors to gather more photons. A laser interferometer gains sensitivity by extending its arms. The expansion and contraction of space-time are simply easier to discern as longer and longer distances are fully measured because the effect is cumulative. If the mirrors are twice as far apart, they will move twice as much relative to each other when a gravity wave passes by. Forward imagined that eventually two instruments should be built on opposite sides of the country. But by the end of the tests at Hughes, he had exhausted the funds his company was willing to spend on a gravity wave telescope. Chapman had left the astronaut corps in 1972, which meant funding from NASA to expand his prototype was not available either. Consequently, the Hughes Laser Interferometer Gravitational Radiation Antenna project came to an end. But others would continue the advancement of this new approach. One of the most innovative was Glasgow physicist Ron Drever.

Drever's interest in gravity was sparked around 1959, just a few years after he received his Ph.D. in nuclear physics from the University of Glasgow. He came up with an intriguing way to test Mach's principle, the suggestion made by Ernst Mach that inertia, the tendency of an object to resist acceleration, arises when a mass interacts with all the other masses in the universe. With that hypothesis in mind, it was plausible to assume that a particle would accelerate differently toward a large collection of mass, such as the center of our galaxy, than at right angles to it, in a direction where mass is more sparse. That's what Drever tested. The particle in his case was a nucleus of lithium. When excited by a magnetic field (for this test the natural magnetic field of the Earth), the lithium could be made to produce an electrical signal at a specific frequency, a distinct spectral line. "I watched that line over a 24-hour period as the Earth rotated. As the axis of the field swung past the center of the galaxy and other directions, I looked for a change," recalls Drever. A change would indicate that the lithium was indeed being accelerated differently, depending on whether it was directed toward or away from the massive galactic center.

Others had done similar experiments, but Drever, like so many physicists a gadget lover since childhood, did it in a very offbeat way. He put together car batteries and assorted odds and ends in his backyard garden and ran the test from there. It was hardly makeshift, though: his experiment could have detected a shift nearly as small as one part in a trillion trillion. "It beat everyone else who was trying to do it with much fancier stuff," says Drever. In the end he detected no change at all, at least to the level he could measure. Inertia seems to be the same throughout the universe, no matter where a mass is headed. Such tests in physics are now known as Hughes-Drever experiments. Yale physicist Vernon Hughes independently conducted a similar test at the same time. Afterward, Drever spent a year at Harvard, where he constructed sensitive radiation detectors for Robert Pound's gravitational redshift experiments.

Through the 1960s Drever built detectors for nuclear physics and other applications. He also dabbled in cosmic-ray physics, studying the light emitted as the cosmic particles raced through the atmosphere. During a visit to southern England to conduct these tests, Drever stopped by Oxford University to hear Joe Weber lecture about his recent claim to have discovered gravity waves. Drever immediately thought, "If he's right, I'm sure we can do better than that," which brought him into the infant field. He and his group in Glasgow eventually got two bar detectors operating but ultimately saw nothing "Weber was wrong, and I was very sad," says Drever. "I was hoping he was right, because then we'd be in business."

Having no experience in cryogenics, Drever figured he couldn't compete with the supercooled bars then under construction at Stanford and Louisiana State, so he chose a different path. Forward had recently visited Glasgow and had talked with Drever about his pioneering experiment in the Malibu basement. "I thought that the interferometers were likely to be better in the end and also much cheaper," says Drever, a prime consideration in Scotland, where funds for novel projects were scarce. One staff member was good at hunting down local companies that could make things cheaply. The vacuum tanks for their first bars had been made by a firm that manufactured ovens and other food industry equipment. Using old bar equipment and secondhand parts, Drever's group constructed its first interferometer in 1976. The one expensive item was the laser.

Drever quickly learned that laser interferometry was going to be far more difficult than he had initially imagined. The first problem that cropped up was simply light scattering. As light bounced back and forth between the mirrors in the interferometer, with each reflection tracing its own glowing thread as it hit a different part of a mirror, much of that light got lost. It scattered off imperfections in the mirrors. Drever's solution was to switch from a Michelson interferometer to a Fabry-Perot interferometer, a scheme that allowed the light over its many round-trips to stay as one beam and confine its reflections to a small area of each mirror. This reduced the chance that the light might ricochet off a "bump" on the mirror and go off in the wrong direction to ruin the measurement. It also increased the light efficiency tremendously. "The big advantage to me as a Scotsman was that this design was much cheaper," says Drever with a wry smile. That was because the mirrors in this case could be made smaller, as well as the vacuum pipes. That was decisive for the field's advance. The technology did not yet exist to polish particularly big mirrors to the fine levels of smoothness required in this endeavor.

But there was also a downside to this new design: it would not work unless the laser was extremely stable, far steadier than any laser yet available. At that time the wavelength of light put out by the laser would sporadically jitter, which would have made it impossible to keep track of the infinitesimal shivers in the masses mounted in a gravity wave detector. Undeterred, Drever simply invented a means of keeping the wavelength of a laser's light pure and steady. He later discovered that the idea was similar to one Robert Pound had used earlier for microwave cavities. Drever figured a laser could be stabilized—its frequency kept fixed—through a feedback mechanism, locking the laser in a special way to an optical cavity. He visited John Hall, a leading laser expert at the Joint Institute for Laboratory Astrophysics in Colorado, to build such a stabilized laser, since the institute had ready access to the necessary parts. Drever and his colleague James Hough also built a cruder version in Glasgow. "The Glasgow one was comical," says Drever. "It was largely contained in tobacco tins. At that time Jim used to smoke a pipe a lot, and so he had dozens of tins around.

They made good screening for the circuitry. The device used about a dozen tobacco tins."

Just as Drever in Glasgow, Weiss at MIT, and a seminal group in Germany were beginning these investigations into laser interferometry, Kip Thorne was working on the theoretical end of this enterprise. He was making the theory of general relativity "user-friendly" in the search for gravity waves. It was the time when Thorne was a rising star at Caltech, working with his students in making general relativity more testable by generating sets of parameters that experimenters could measure. In the process Caltech was replacing Princeton as a world center in relativity theory. Thorne himself became involved with gravity waves in 1968 when he was introduced to Braginsky. Immediately impressed by the Soviet researcher, Thorne began a collaboration. Until the end of the Cold War, Thorne would spend about a month in Moscow every other year, becoming the unofficial "house theorist" for Braginsky's gravity wave group. From Braginsky, Thorne became convinced of the long-term experimental possibilities of gravity wave detection, although he was skeptical of any short-term success. His wary attitude would soon change, though.

The turning point came at a meeting in the medieval town of Erice, Sicily, a favorite summer spot for physicists to convene at the Ettore Majorana Center for Scientific Culture. Held in an old monastery perched on a cliff overlooking the Tyrrhenian Sea, the 1975 conference had been organized by Weber to take stock of the field and discuss advanced techniques. As a theorist, Thorne had been calculating the gravity wave strengths expected from various astronomical sources. Listening to the experimental presentations, he began to see that researchers had a good chance of getting to the required sensitivities with advanced techniques, such as the use of special materials or going to low temperatures. "I came away from that meeting convinced that the field was very likely to succeed," he says. "I hadn't had that kind of conviction before. It was because I was seeing the ideas for improvements in the detectors and what one might plausibly expect those ideas to achieve over 10- or 20-year timescales." As a result, Thorne became the field's most dedicated barnstormer, going around the United States giving talks about the field's promise and the sources that might be detected with the new technologies coming online. "The Weber controversies had left a black mark on the field to some degree," notes Thorne, "and there was the need to erase that and maintain momentum in the United States."

Thorne was instrumental in convincing the Caltech faculty and administration to establish a gravity wave detection team at the university, a natural complement to his theory group. Thorne was not wedded to any specific approach at first. "My attitude," he says, "was to leave it up to whomever we hired to decide the best direction." Weiss, though, hoped to change his mind. Weiss was chairing a NASA committee in 1975 on relativity experiments that NASA might carry out in space. He was already chairman of the science working group on another big NASA project, COBE, the Cosmic Background Explorer satellite then being built to measure the vestigial hiss of the Big Bang with exquisite precision. Along with his gravity wave investigations, Weiss has been a major player in ongoing measurements of the microwave background, first with balloons and then from space. He was one of COBE's originators. Because of Thorne's expertise in general relativity, Weiss invited Thorne to come to Washington and speak to his NASA committee. That evening in their hotel they stayed up nearly all night in conversation about gravity wave detection. Thorne at the time didn't hold out much hope for laser interferometers. In a section on gravity wave detectors in his book *Gravitation,* Thorne had written, "As shown in exercise 37.7, such [laser interferometer] detectors have so low a sensitivity that they are of little experimental interest." That night Weiss would begin to convince Thorne that laser interferometry was a contender. (Today, Weiss keeps a copy of that quotation posted on his office door, just to tease Thorne whenever he visits MIT.)

After a committee study, Caltech agreed with Thorne to recruit the world's leading expert in gravity wave detection, someone who would direct construction of a sophisticated prototype, either bar or interferometer, that would allow the university to refine the techniques and hardware necessary for a future gravity wave observatory.

Thorne would have liked to have brought in Braginsky, but with the Cold War still in progress it was not possible. Braginsky feared the consequences to his family and colleagues. The transfer would have been viewed as a defection. Weiss was then heavily involved in the early phases of COBE, which diverted his attention. But another name was often at the top of the lists of Thorne's consultants: Ron Drever. "Drever then had the best track record in terms of dealing with technical obstacles. He was preeminent. He had beautiful ideas, ideas that people would pooh-pooh at first and now they're incorporated into LIGO," says Thorne. Drever's research team in Glasgow was just then beginning to improve its laser interferometer, employing Drever's latest modifications. The size of the instrument was fixed by the length of the available room, an old particle accelerator laboratory. The interferometer arms were 10 meters long. "It was a struggle to make it all work," says Drever. It was in the midst of this start-up that Caltech began to vigorously court Drever, who was ambivalent about moving just as his Glasgow group was making progress. Although Caltech was the "big leagues," as Drever puts it, he was more attracted to the way in which European universities supported new endeavors. "I was quite happy where I was. You could do a lot with little money. The university employed technicians who could be used on any project. That meant you could try out new ideas without having it tied to some grant," notes Drever. By 1979, though, Drever finally decided to spend half of his time in California, which gave both him and the university the chance to see if the new Caltech venture was viable. It was. After five years Drever became a full-time faculty member. And with Drever on board, laser interferometry became Caltech's method of choice.

The growing momentum of laser interferometry had already caught the eye of the National Science Foundation. When Richard Isaacson arrived at NSF in 1973 as associate program director for theoretical physics, he recalls his predecessor giving him a bit of parting advice: "I was visited a few weeks ago by a very clever guy—Rai Weiss," said Harry Zapolsky. "He has some interesting new ideas about gravity wave detectors. If he comes back, you should pay attention." Eventually reviewing its national program of gravitational wave detection at the end of the 1970s, the agency decided to expand its funding into this new arena. Also influential was Caltech's stepping forward and investing its own money in the technique. "Physicists tend to follow one another," notes Weiss, who had faced delays when he first approached NSF for funding. "Once a large and prestigious university decided to go into it, that gave an extra little nudge."

With money from both Caltech and NSF, Drever proceeded to set up a full-fledged gravity wave laboratory on the northeast corner of the university's campus. Stan Whitcomb, who became Drever's right-hand man, was brought on board to assist with overseeing its construction. Caltech's aim was to build an interferometer identical to the one in Glasgow, only bigger. This interferometer now resides in a one-story structure, unassuming in its beige tone, that wraps around a corner of the university's engineering shop, forming two long corridors. Only a modest sign on the door reveals the building's purpose. Inside, the lab's most prominent features are two 40-meter-long steel pipes meeting at right angles. The 40-meter length was not chosen for scientific considerations. Drever would have gone even longer, but a tree was in the way, a tree that no one was eager to cut down. A vacuum chamber stands at the corner of the L as well as at each end of an arm. In each chamber a mirror/test mass is suspended. Each mass is a 5-pound cylinder of fused silica. (When first built, the masses were mounted in glass tanks christened Huey, Dewey, and Louie, after Donald Duck's nephews—a tip of the hat to nearby Disneyland and typical of the school's humor.) Inside the long pipes the laser beams reflect back and forth—you might say from Dewey to Huey and from Dewey to Louie. Vacuum pumps silently work in the background, keeping the pipes evacuated from stray atoms that could disrupt the light's journey.

To protect the suspended mirrors from such outside disturbances as passing trucks or the seismic tremors that occasionally shake Pasadena, the supports from which the mirrors are hung are cushioned by layers of stainless steel and rubber. When first set up in the early 1980s, toy cars were used for the cushion, a colorful assortment of tiny pink, green, yellow, red, and blue rubbery sedans. It was a clever and cheap rubber source at the time but turned quite troublesome in the end. Outgassing from the toys dirtied the vacuum system. Starting in the early 1990s, the Caltech prototype was completely refurbished, with a new vacuum system and new pipes. The goal was to make it a smaller version of the full-scale observatories then planned for Louisiana and Washington. It is now a test bed for future innovations. A graph on the wall depicts its evolution. When first operating in the early 1980s, the prototype reached a "modest" strain of 10^{-15} (able to discern a movement as small as an atomic particle). By 1994 the system reached a strain of 10^{-18}, a thousandfold improvement. The progress was largely due to a slow but continuing series of technological improvements. Laser power, for example, has increased over the past 20 years, which directly affects the sensitivity of a laser interferometer system. The equipment is also better isolated from seismic disturbances. And, perhaps most crucially, the Caltech detector now uses "supermirrors" as its test masses. Made of layers of dielectric material, they lose only 100 photons for every million reflected.

As soon as he arrived in California, Drever started checking out all the commercial vendors to find out who was making the best mirrors. He heard that Litton was making special mirrors for the military for use in laser gyroscopes. They were not as yet available commercially, but Drever forged a connection and arranged for the company to make a special batch for his new interferometer. "They were fantastic, at least a 100 times better in terms of reflectivity losses," he says. With such mirrors in hand, Drever was spurred to think of additional improvements. When checking out his new supermirrors, he noticed that the reflected light was strong enough to bounce within the interferometer many times. Given such low light losses, he figured he could "recycle" the light, have it bounce back and forth over and over again, which essentially boosts the power of the laser and increases the instrument's sensitivity. It was the development of the "supermirrors," with their miniscule losses, which allowed Drever to even consider such power recycling. At the time this idea was revolutionary. "You'd catch the light and send it back in," says Drever. Now it's standard practice. When a visible-light laser is used, the effect can be spectacular. The laser beam enters the detector and reflects back and forth between the mirrors in each arm some 100 times. It's as if 100 laser beams are superimposed on one another. If the mirrors are aligned exactly right, so that the beams are in phase, the relatively dim laser beam suddenly brightens within the cavity into a brilliant shaft of light. With those two key improvements—a stabilized laser and power recycling—Drever enabled laser interferometry to turn a corner as a gravitational wave detection system. The technique looked more and more promising in its ability to reach the sensitivities needed to conduct astronomical investigations.

The Caltech prototype has never been a true gravity wave telescope, more a working model to continually improve the instrument design. But that didn't keep Caltech investigators from attempting a test. For 12 days and nights in the winter of 1983, the Caltech interferometer was hastily put on the air after radio astronomers discovered a neutron star spinning what was then a record 642 revolutions per second, possibly jiggling space-time in the process. The 1987 Magellanic supernova was examined, too, though days after the initial burst. In both cases the Caltech detector perceived not a wiggle.

From his previous experience as an experimentalist, Weiss had initially envisioned the field of gravity wave astronomy growing steadily but very, very slowly. He figured that researchers in this arena would have to work on the innumerable technical challenges of laser interferometry before attempting a full-blown observatory. But a series of pressures and frustrations soon changed his assessment. Several years after Weiss returned to MIT's Research Laboratory for Electronics upon completing his postdoctoral fellowship at Princeton, its mission changed. Previously, federal grants to the lab could be applied to whatever ideas the lab was currently working on. Indeed, such funds helped him set up his first laser interferometer, a prototype with 1.5-meter-long arms, and support the graduate students building it. But in the thick of the Vietnam War, a new rule was imposed which required that all research funded by the Defense Department, the lab's major funder, have a direct bearing on the military's needs. As a result, cosmological and gravity-related projects eventually lost their support. At the same time, Weiss was getting little respect from the MIT physics department, which was then more concerned with enhancing its solid-state division. "The faculty gave my students such a terribly hard time," he says. "They sneered at the extremely low sensitivity of the instrument." Indeed, in those early days Weiss found it difficult to convince many in the physics community that this new approach to gravity wave detection had the potential to surpass the bars in sensitivity. Some thought the scheme was far too complicated—and perhaps even an erroneous method of detection.

As a last-ditch effort, Weiss arranged with the city of Cambridge to have the road right outside his laboratory—Vassar Street—closed down two weekends in a row at night. Stopping the trucks from rattling down the road would give his students the necessary peace and quiet to run their sensitive tests. Sawhorses were set up in the street to block the traffic. "We were trying to get my students a thesis," says Weiss. They obtained a strain of 10^{-14}, decent for a small prototype but impossibly weak for astronomical investigations. "During their oral exams," continues Weiss, "my colleagues had the temerity to ask these kids what they discovered. 'Well,' said one student, 'We didn't see the Sun blow up.' One professor replied, 'I can look out the window to see that. What do we need your data for?' They didn't see the technological aspects of it at all. That's when I decided I was never ever going to put a student in that situation again." With his funding slashed and his colleagues indifferent, Weiss realized he had to get a full-scale observatory under way. He needed to get into the business of astrophysics as soon as possible. And that meant going beyond tabletop detectors or prototypes with extended arms. It meant going big—very big. In 1976 he began to work on an idea that would become the seed of LIGO.

Right off Weiss envisioned a system with two widely separated detectors. He initially set the length of the arms at 10 kilometers, a little over 6 miles. Such a long length was necessary to make the instrument sensitive enough to detect the waves that theorists said were out there. Knowing that such a sizable facility would cost tens of millions of dollars, Weiss reasoned he'd have to call an international meeting and try to turn its construction into a worldwide effort. He never imagined he could obtain funding from the National Science Foundation alone, by then the only U.S. funder for gravity-related research. "I knew right off that this was going to cost over $50 million," says Weiss. "It was outrageous to contemplate that the NSF would go into this field that (a) had that terrible history with Weber, (b) had no experience with big projects, and (c) had no scientific basis as yet. It was cuckoo."

But as soon as NSF made its decision to fund the Caltech prototype, a sizable investment for the agency, Weiss decided to be more aggressive and offered his idea to leapfrog to an even bigger detector. Among gravity wave researchers at the time, Weiss was the only person with a firm grounding in major physics projects, being in the thick of building the COBE satellite. From that experience he came to learn it would be wise to build a large facility fairly quickly, and then develop and advance the techniques along the way. "Rai recognized, far more clearly than anyone else, that the only way to get to the required sensitivity was to build something with long arms," recalls Thorne. NSF gave Weiss the go-ahead to conduct a feasibility study of his ambitious scheme, then called the "Long Baseline Gravitational Wave Antenna System." Most needed were realistic cost estimates for constructing such a facility. Completed in 1983, in collaboration with Peter Saulson and Paul Linsay, the study (now familiarly known as the "Blue Book" for the color of its cover) ultimately convinced NSF to initiate research and development. The decision came with one understanding: any major laser interferometer observatory had to be a joint project between Caltech and MIT. There were both political reasons—the combined clout of two prestigious institutions to get the financial support from Congress—and technical reasons. Big interferometers required a much larger effort than you could mount with one professor and a handful of assistants. Weiss had anticipated such a collaboration forming all along.

With this decision, gravity wave astronomy moved toward the big leagues. Weiss, Drever, and Thorne were in charge of overseeing the new collaboration. Given Thorne's Russian connection, the three came to be known as the "troika." The troika's first order of business was to put together a detailed construction plan for two full-scale interferometers, which had been reduced to having 4-kilometer arms due to budget and engineering considerations. (The available sites, for one, were limited in size.) The NSF's response to their initial ideas, though, was decidedly cool. Their suggested schemes were judged not good enough to be viable, especially in an era when federal budget woes were putting the brakes on other big science projects. "Caltech and MIT were simply not ready," says NSF's Isaacson. "Their plans were premature." As a result, the troika went back to the drawing board to rework their plans but continued to get lukewarm critiques. Normally, that would have been the death knell for a proposed science endeavor, but Isaacson and Marcel Bardon, then director of NSF's division of physics, had faith in the idea and pulled every string to keep the project alive. They made sure that money was provided to continue research and development. "Bardon in particular recognized the technological promise," recalls Isaacson. "The intellectual excitement was overwhelming. But what we had to do was reduce the risks. We wanted to nurture the dream until Caltech and MIT were up to the job of managing such a project."

"It was a miracle," says Weiss. "So many things would have killed it, but the National Science Foundation was responsible for keeping it going." As a result of this favored handling, the proposal became highly visible within the scientific community—attention that brought much criticism along with it. Richard Garwin, Weber's nemesis, began to loudly question the worth of building a large gravity wave observatory so soon. He did not trust the grand claims being made for it by its supporters. With assistance from Caltech and MIT, NSF answered by assembling a blue-ribbon panel, including Garwin, to advise the agency on "going big." The panel met for a week in Cambridge, Massachusetts, in the fall of 1986, bringing in the major players in gravity wave detection from around the world to discuss the prospect. Also at the meeting were members from industry to discuss the technical feasibility of making the required optics, lasers, and servo systems. "It was the turning point for the field," says Thorne. "This hard-nosed committee produced in the end, after much internal debate, a unanimous report. It said that the field had great promise and the correct way to do it *was* to build two big interferometers right from the beginning, because you can't get any science with one by itself."

But this support came with a strong caveat: urged on by Weiss and others, the committee recommended that the troika be disbanded and replaced by one project director. There had been disruptive tensions between Weiss and Drever all along, technical differences of opinion that made it difficult for the two institutions to work together effectively. "It was five years of sheer agony for everyone," says Weiss. Drever has always been driven by an intuitive physical instinct. Weiss is far more analytical. Drever is essentially a loner, while Weiss is experienced on big projects and more realistic about the compromises required in such a setting. Drever preferred fine-tuning the prototype before jumping up in size. Weiss was eager to build big, then tweak. Thorne was caught in the middle. This mismatch in temperaments became a serious problem for the project, holding up some critical decisions. Consequently, NSF demanded that a single director be put in place with authority over the entire project. Caltech and MIT found that director in Rochus Vogt, known to all as Robbie.

Vogt took over in June 1987. He had a distinguished track record. Trained in cosmic-ray physics at the University of Chicago in the 1950s, Vogt served as the first chief scientist at the Jet Propulsion Laboratory in the mid-1970s, making science preeminent at the NASA facility. "He was also one of the best chairmen I have ever seen heading up the division of physics, math, and astronomy at Caltech," says Thorne. It was then that Caltech was building its millimeter-wave radio astronomy array in California's Owens Valley, which was in dire trouble during its construction and about to be canceled. Vogt went in and pulled it together on time and on budget. But Vogt is also known as a tough man, who garnered political enemies over his career. As a youth in Nazi Germany, he developed a fervid distaste for wasteful authoritarian bureaucracies. With his short-clipped hair and black-framed glasses, he strikes one as a taller and leaner Henry Kissinger. Vogt was available only because he had just been fired as

Caltech's provost, due to conflicts with the university president. Reluctant to accept the directorship at first—he yearned to get back to being a "real scientist who plots data"—Vogt eventually relented. University trustees sold the project to him as Caltech's next Palomar telescope, for many years the most powerful optical telescope in the world. Vogt's organizational skills turned out to be invaluable to the project, which at last had a name—LIGO. Vogt personally husbanded the final proposal, getting his fingers into all the nitty-gritty details and corralling everyone to get involved. He thought of himself as the "resident psychologist." Right off, he broke the scientific deadlock between Drever and Weiss, by choosing Drever's Fabry-Perot design over the Michelson interferometer favored by Weiss. He even persuaded Thorne to turn experimentalist and come up with a successful solution to extinguishing stray light in the vacuum pipes. Vogt, now LIGO's most ardent champion, was certain that if the project were terminated, it would kill gravity wave astronomy for a generation.

The final proposal for LIGO got strong reviews at NSF in 1990. An external review panel also gave it a thumbs up. But the requested money was so sizable—an initial $47 million outlay of a total $211 million construction cost—that it had to get Congress's okay. This was a first for NSF. Unlike the Department of Energy, which regularly deals with large projects such as particle accelerators, NSF had never before sponsored a project so large that it required line-item approval in the federal budget. Opposition immediately arose in the astronomical community, which proclaimed that such money would be better spent on telescopes. At the time $211 million was twice the total of the NSF astronomy budget. Astronomers were angered that NSF had chosen to put the money into a gamble, rather than into already-proven technologies. Was it worth such a high price to find a gravity wave, the critics were asking Congress? Nobel laureate Philip Anderson, a condensed-matter physicist, wondered aloud, "If it didn't have Einstein's name on it, would you give a damn?"

LIGO researchers were particularly dismayed when a former member of the gravitational wave detection community, Tony Tyson, testified against the project before a panel of the House of Representatives Committee on Science, Space, and Technology. He stressed to Congress that LIGO "demands a truly phenomenal increase of sensitivity," up to 100,000 times more sensitivity than the Caltech prototype was then currently attaining, before it could hope to obtain important astronomical data. A LIGO endeavor, to Tyson, was simply "premature," since too many engineering problems were still unsolved at the time. He preferred a slow-but-sure approach. "Innovation is not necessarily synonymous with a 'shot in the dark,' " he concluded.

Berkeley astrophysicist Joseph Silk got a jab in while reviewing a popular book Thorne had written on general relativity. Silk wrote that LIGO "has misleadingly cloaked itself as an 'observatory.' Despite its name, any astronomy on the symphony of waves from merging black holes in remote galactic nuclei will have to await a greatly refined second-generation detector, and undoubtedly a vastly more expensive undertaking." Even bar detector scientists liked to point out that bars were cheaper and more developed. Others questioned the importance of gravity wave studies altogether, insisting that scarce government funds would be better spent on projects with lower price tags and far likelier scientific payoffs. LIGO supporters countered that they were after more than sheer detection. They resented it being called a shot in the dark. What they wanted, they said, was to open up a whole new arena for gleaning information from the universe, a method far different than gathering electromagnetic radiation. Including the word "observatory" in LIGO's title was a deliberate choice, an expression of their intention to use LIGO as an ongoing experiment. Moreover, they noted, LIGO's construction funds were independent of the NSF's regular budget. Rejection of LIGO did not mean the funds would necessarily be applied to other science projects.

Even with blue-ribbon panels strongly supporting the project, Congress became wary when eminent scientists stepped forward to object. LIGO was stalled for two years. Vogt was a relative novice in his first dealings with Congress. In 1991 he failed to get the go-ahead for construction, although he did get funding for further engineering and design work. Concerns over the federal budget deficit were high. Congressmen questioned whether they were ready to invest such a large sum of money in an unproven facility. The next year Vogt honed his lobbying skills with the help of a consultant, learning to sell the story of gravitation to key legislators. While in Washington, for example, Vogt wrangled a short 20-minute meeting with Louisiana Senator J. Bennett Johnston, who later became an ardent, behind-the-scenes supporter of LIGO, especially when his home state was chosen as one of the two sites. "After I had my 20 minutes," recalls Vogt, "Johnston's senior staffer looked at his watch and said, 'Senator, the 20 minutes are up. Let's go.' But the senator responded, 'Cancel the next appointment.'" Johnston was so captivated by Vogt's tales of cosmology that he cancelled the following appointment too, as well as the one after that. The two ended up sitting on the floor by the coffee table, while Vogt drew pictures of curved space-time. Einstein's name once again wielded its magic power. In the end Congress appropriated the money.

With the funding came a sudden shift in the tenor of the project. The transition is quite evident at both MIT and Caltech. Blueprints and photographs line the hallways. Memos clutter the desktops. The offices more resemble an industrial corporation than an ivory tower. The initial laser interferometer team, when it first started up, was quite small. There were about a dozen people on each coast, including technicians, scientists, engineers, and administrative personnel. Today, there are more than 150 on staff, two-thirds at the Caltech headquarters and the rest at MIT and the two detector sites. The jump from prototype to mature facility was a canyon-sized leap: the arms going from 40 to 4,000 meters, a factor of 100. "It was a big transition," says MIT's David Shoemaker, a deputy detector group leader for LIGO. Scientists had to move from their individual lab environments, where they had total control, to a hierarchical facility. Participants now document their every move and deal with myriad outside companies. Once a singular pursuit, gravity wave detection has become a networking of many players, each serving a specific role.

There were repercussions to this change. As with any burgeoning enterprise in science, LIGO has had its share of heated discussions, compromises (both political and scientific), and struggles between pioneering scientists with strong and volatile personalities. New fields often attract the risk takers, whose passions and fervor can be difficult to handle day after day. Some observers even label it hubris, the unwillingness to concede that someone else might see a better way to handle a particular problem. "Most of the conflicts could be blamed on the shift from tabletop physics to big physics," says Peter Saulson, who has been a LIGO participant since its beginning, now as an independent researcher. "It's a time when you have to step out of your laboratory, set up deadlines, and follow budgets. Many of the original people didn't have this experience. Knowledge had to be brought in from the outside. At first we thought we were uniquely cursed. But I have since found out that it's part of every major science endeavor, where you invent something from scratch and transfer it to a larger arena." Construction of the 200-inch Palomar telescope, for decades the largest optical telescope in the world, faced an identical crisis. Ronald Florence, writing on the telescope's long development, said that the scientists' "insistence on exploring, designing, and engineering every step of the project from scratch—doing basic research on subjects as varied as oil bearings, the wind resistance of dome sections, and the chemistry of glass—was for [the chief administrator] a sure route to a quagmire of indecision that would never see the telescope built. . . . Men who were building a unique machine, an instrument perfect enough to explore the secrets of the universe, didn't like that attitude."

A LIGO member with experience in industry saw this history play out once again. Research scientists, long used to laboratory independence—the freedom to change an experiment at will—became upset over LIGO's rigorous schedules and their inability to make last-minute changes in the instrument. Scientific considerations suddenly had to bow to financial and engineering concerns. This meant that certain technologies had to be locked in early, even though advances may have developed later. Some adapted; others left. One unwilling victim was Ron Drever.

Originally brought in to jump-start Caltech's entrance into gravity wave detection, Drever was unceremoniously ousted in 1992 after

PSR *1913+16B*

RA	DEC	PERIOD	DM
19.13.74	16°00'	.059/0332	150
191313	16°00'24"		167±5
±4ˢ	±60"	58 98/24	
		.059030	

SCAN	DUMPS	DM CHNL	FREQ/tree	S/N
473	705-936	8,10	430/2B	7.25
526	1-128	8,9	430/2B	13.0
535	1-384	8	430/2B	15.0

$\ell = 49.95$ $b = 2.11$ *fantastic!*

Russell Hulse (right) shown in 1974 operating his computer and teletype at Arecibo observatory in Puerto Rico. The form records the "fantastic" detection of PSR 1913+16, with its ever-changing periods scratched out by Hulse in frustration. (© The Nobel Foundation)

Joseph Taylor found evidence for gravity waves in the motions of PSR 1913+16, a pair of neutron stars. (Rita Nannini)

Joseph Weber (right) founded the field of gravitational wave astronomy four decades ago with his invention of the bar detector. His reported detections, though, are still a matter of great dispute. He is seen here in the early 1970s working on one of his bars at the University of Maryland. (courtesy of Joseph Weber)

John Archibald Wheeler brought general relativity to the forefront of astrophysics and gave the black hole its name. (photo by Robert A. Matthews; courtesy of Princeton University)

In 1887 Albert Michelson built this interferometer in his basement lab and failed to detect the ether, a perplexing finding that was ultimately resolved with Einstein's theory of special relativity. (Carnegie Observatories Collection, Huntington Library, San Marino, California)

MIT's Building 20 was the famous "plywood palace," where in the 1970s Rainer Weiss first devised the plans for a laser interferometer that became the seed for LIGO. (The MIT Museum)

Caltech theorist Kip Thorne (seen here in the 1980s) became gravitational wave astronomy's most vocal advocate once he was convinced that detection of a wave was within reach. He and his students have specialized in predicting the possible celestial sources. (courtesy of Robert J. Paz/Caltech)

Ronald Drever developed several engineering innovations in Scotland and at Caltech that enabled laser interferometry to go forward. (© James Sugar)

LIGO researcher Peter Saulson of Syracuse University: "The levels of precision we are striving for mark our business; if you do this, you have the 'right stuff.'" (Peter Finger)

Princeton's Robert Dicke, seen here in the 1960s, was a key figure in the resurgence of general relativity experiments in the second half of the twentieth century. (Princeton University Library)

James Faller, Peter Bender, and R. Tucker Stebbins (left to right) initiated research on a space-based laser interferometer at the University of Colorado's Joint Institute for Laboratory Astrophysics. Their work ultimately led to the proposed LISA endeavor. (photo by Ken Abbott; © University of Colorado)

All the mass concentrated in the galaxy cluster Abell 2218 acts like a giant zoom lens. Located 1 to 2 billion light-years away, the cluster amplifies and bends the light of more distant galaxies into concentric arcs, just as Einstein's theory of general relativity predicts. (NASA: Andrew Fruchter and the ERO team)

On the left, LIGO director Barry Barish. (FermiNews) Above, MIT's Rainer Weiss, laser interferometry pioneer and for many the founding father of LIGO. (Caltech)

Members of the VIRGO team pose in front of their advanced seismic isolation tower, the "super attenuator." VIRGO codirector Adalberto Giazotto is in the back row, third from the left. (courtesy of VIRGO)

In clean-room attire, LIGO researchers set up one of the many optical support systems within the vast interferometer. (Caltech)

The Laser Interferometer Gravitational-wave Observatory (LIGO) situated in the pine forests of Livingston, Louisiana. One of its 4-kilometer-long arms runs upward, the other to the right. A duplicate observatory resides in Hanford, Washington. (Caltech)

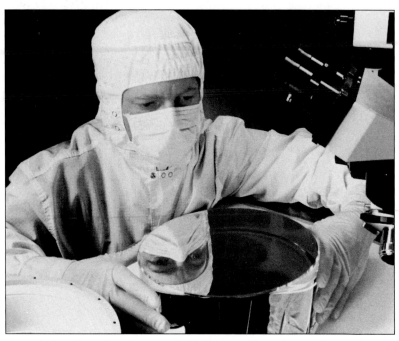

An optical engineer handles one of LIGO's pure silica mirrors, the very heart of the gravity-wave detector. (Caltech)

Members of the VIRGO team pose in front of their advanced seismic isolation tower, the "super attenuator." VIRGO codirector Adalberto Giazotto is in the back row, third from the left. (courtesy of VIRGO)

In clean-room attire, LIGO researchers set up one of the many optical support systems within the vast interferometer. (Caltech)

The Laser Interferometer Gravitational-wave Observatory (LIGO) situated in the pine forests of Livingston, Louisiana. One of its 4-kilometer-long arms runs upward, the other to the right. A duplicate observatory resides in Hanford, Washington. (Caltech)

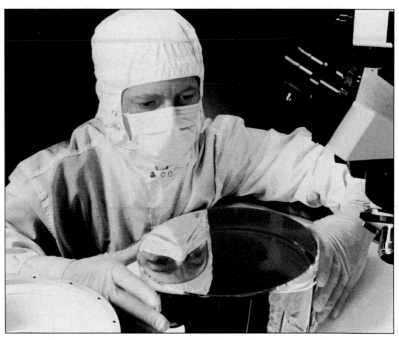

An optical engineer handles one of LIGO's pure silica mirrors, the very heart of the gravity-wave detector. (Caltech)

growing conflicts with the LIGO team, particularly Vogt. A short portly man with a bulbous nose and warm blue eyes—a Santa Claus without the beard—Drever is always talking and chuckling, ideas constantly bubbling up fast and furiously. And for some this rambling creativity posed a problem. Drever was most comfortable working in his own laboratory, where he could constantly alter his experiments as he developed new schemes. He was reluctant to lock into a final design, if a better method was on the horizon. Drever was a veritable fountain of ideas, as many bad as good, but he would champion them all with equal energy.

And then there was Vogt, whose management skills played a large part in getting LIGO's final approval. He was a man revered by many at Caltech for his organizational talents yet who was also feared for his sporadic outbursts and at times harsh tactics. He was in battle mode as he fought both a skeptical scientific community and a wary Congress for acceptance of LIGO in the early 1990s. You were either friend or foe to the LIGO venture, an uncompromising stance that some found hard to deal with. Ultimately, there was a clash of wills: Drever, always coming up with new approaches, wanting to try them out, versus Vogt, the manager wanting to maintain order and discipline because of a rigid time schedule. In a way the strain between them was also a conflict over ownership. Who got credit for LIGO? Drever, the brilliant experimentalist, whose major breakthroughs were recognized worldwide and enabled a facility such as LIGO to go forward, or Vogt, the expert strategist, who got the controversial project approved by Congress?

A review team, brought in to evaluate the LIGO project in the midst of this internal strife in 1992 and 1993, likened the major players to "a dysfunctional family that needs to be split up." (One observer ruefully joked that Caltech needed "to put Prozac in its water coolers.") A university committee eventually agreed that Drever had been inappropriately dismissed, and Caltech provided money for him to continue his research on advanced interferometers independently. Yet even with Drever gone, tensions persisted. Vogt ran a tight ship and kept his decisions close to the vest, for he utilized a managerial approach used at times on covert military projects. "Give me the money and stay out of my way" was his philosophy, a method he had successfully applied in building Caltech's Owens Valley radio telescope array. Vogt preferred working with a small elite team of scientists and engineers, free of governmental control and costly bureaucratic reviews. He was convinced that only then could LIGO be built for $220 million (the planned construction cost was ever rising). But NSF ultimately decided that LIGO needed to be more open and accountable—its construction activities laid out in meticulous detail and a coherent plan developed for accommodating outside scientists. To do this the agency wanted LIGO's management to shift to a managerial mode long used on high-energy physics projects. Vogt resisted.

In response, after lengthy consultation with both MIT and NSF, Caltech replaced Vogt with physicist Barry Barish as LIGO's principal investigator in February 1994, just a month before groundbreaking took place in Washington state for the first of the two LIGO detectors. Vogt, astute in the politics of science, had shepherded the project for seven years from proposal to funded experiment but was less prepared to deal with managing an evolving large-scale facility that was at last being cast in steel and concrete. Barish, though having no experience in gravity wave research, was tapped for his expertise in handling large physics projects. The cancellation of the Superconducting Super Collider in 1993 made Barish available; he had been heading up one of its detector teams. Barish entered particle physics in its golden era during the 1960s, when new particles were being found fast and furiously. It was a time when universities were abandoning their in-house particle accelerators and starting to use large national facilities. Today, along with his directorship of LIGO, Barish oversees an immense detector in Gran Sasso in Italy, which is on the lookout for monopoles, the hypothetical units of magnetic charge. So operating a facility like LIGO is familiar territory for Barish. The monopole search, a joint Italian-American venture, involves large numbers of detectors spread over an area the size of a soccer field. It is located deep underground, to block out disruptive cosmic rays. Barish understands searching for the will-o'-the-wisps of physics. "In a sense the searches for both magnetic monopoles and gravity waves are very similar," he says. "But, theoretically, gravity waves are more solid."

Barish, a longtime member of the Caltech faculty, had watched LIGO undergoing its growing pains from the sidelines. He credits the initial team for performing the necessary conceptual work and pinpointing the technique's limitations. But, he adds, taking the leap from tabletop physics to a miles-long facility was beyond their level of expertise. They had a certain blindness to this limitation, says Barish, even an arrogance when confronted with reviewers' criticism. They were literally founding a field, yet they hadn't grown up in the environment that particle physicists take for granted. The field of particle physics evolved over decades. With LIGO, gravity wave astronomy was growing up virtually overnight. Its evolution from small to big was occurring at a rapid tempo and that meant certain scientific styles had to give way quickly to new approaches, ones more compatible with a large infrastructure. "In a small lab if you make a mistake, you can go in the next day and fix it," he explains. "But here, when you are committed to spending a hundred thousand or a million dollars, you can't fix it later. You need to have a system of checks and balances internally. In particle physics that's just part of the structure." When Barish took over the leadership of LIGO, just months after the Superconducting Super Collider was closed down, he found a "battle-worn" group recovering from the strains of the Vogt-Drever tempest. He set one simple goal: to build LIGO.

A Little Light Music

The Hanford Nuclear Reservation, operated by the U.S. Department of Energy and currently the nation's prime repository for nuclear waste, sprawls over hundreds of square miles of scrub desert within the rain shadow of the Pacific Cascade Mountains in south-central Washington state. John Wheeler first came to this isolated site in the 1940s when the government set up a manufacturing plant in Hanford to produce plutonium, an element that does not occur naturally, for use in the bomb that was dropped on Nagasaki. Settling in nearby Richland at the time, Wheeler remembers "a community of houses, stores, and schools that had been erected in a matter of months by the Army Corps of Engineers. The sidewalks were tar that had been squeezed out like toothpaste. Asparagus sprouted through cracks in the sidewalks from the farm that had been there before the Corps' bulldozers moved in." The town continues to thrive by the banks of the Columbia River.

It is 10 miles from Richland to the Hanford facility, proceeding west along Route 240. The turnoff is not obvious. At the key intersection the road signs direct travelers to either continue west or turn south. There is no sign at all to explain the highway going north, the entrance to the reservation. It is a habit left over from Hanford's many years as a national secret. Five miles down that desolate two-lane road is a duplicate of the LIGO complex in Louisiana—the same cream, blue, and gray colors. Standing alone on the vast plain, a landscape long ago carved flat by the immense outflow of an ancient glacial lake, the observatory resembles either a tasteful warehouse or a modern art museum inexplicably placed in the middle of nowhere. It is a rent-free guest on the Hanford site. The observatory's nearest neighbors, though still miles away, are a nuclear power plant and a moth-balled research reactor. They share a vast, cloud-studded sky that stretches from horizon to horizon. Only to the southwest is this blue vault interrupted by the Horse Heaven Hills and the smoothly sculpted Rattlesnake Mountain.

Tumbleweeds, the dense Russian thistles accidentally introduced to the west, are ubiquitous. They continually roll over the barren terrain and pile up along the arms of the interferometer. "We baled 200 tons of it last year," says Fred Raab, head of the Hanford observatory. "We have a crew out every week." The strawlike balls are gathered and baled like hay and then used for erosion control around the site. Otherwise, the roads along the miles-long arms would be completely blocked.

Raab enjoys his job immensely. "Being first, you get to write the playbook," he notes. Raab joined LIGO when he was convinced that there would be gravity wave detections in his lifetime. Considering his background, it's the ultimate challenge. Trained in atomic physics, he has long dealt with precision measurements on the tiniest of scales. "To me a table is a bowl of Jell-O," he says, a reference to the constant jitters that occur on the atomic level. The gravitational jitters he seeks with LIGO, though, will be even smaller.

The journey begins in the complex's center station, an enclosed city of metal gleaming under bright fluorescent light. One and a half million pounds of metal were used to construct the instrument. Seven hundred truckloads of concrete were brought in to build the floors and beam-tube roadways. The vacuum chambers, where the test masses reside, look like the tanks in a microbrewery, though unpolished. "It's more like a sewage treatment plant," says Raab with a laugh. There is no noise, however. Only a whisper of wind from the air ducts. Gravity wave detection starts with the laser, set in an alcove of the main hall. The laser sends its light into the vacuum system, where it is divided into two beams, each directed into a separate 4-kilometer arm. As in Louisiana, the arms shoot outward to form the familiar L shape. In Hanford one arm is directed toward the northwest, the other to the southwest.

Given Hanford's legacy of covert operations, some locals harbor the suspicion that the observatory is just a cover-up for a secret laser weapon project. They imagine the arms are somehow hinged, able to rise up and fire a powerful ray blast into space. Raab, an enthusiast for outreach education, built a small tabletop interferometer to take to local schools to demonstrate what is really happening in those miles-long tubes. Its light source is a toy laser pointer, the kind often used by lecturers. It's held up by a clothespin. The vivid red light is directed into a "beam splitter," which creates two separate beams that are sent off at right angles. Each beam reflects off a mirror. On their return, the two beams are recombined and aimed at a white screen. When the two beams are "in phase," wave peak matching wave peak, a bright red spot is seen on the screen. When the two beams are "out of phase," the trough of one wave canceling the peak of the other (like +1 and −1 summing up to zero), a dark spot is seen. By pulling on a string to move a mirror ever so slightly, which changes the distance in one of the arms, the spot changes from bright to dark or dark to bright. The kids take delight in this effect. It is exactly what is happening inside LIGO. A tiny change of distance in an arm translates into a change of light intensity at the interferometer's output.

LIGO is a direct descendent of the Michelson-Morley experiment, whose failure to detect any variation in the speed of light led to the theory of relativity. But where a detectable change in that experiment would have been relativity's downfall, a change in the LIGO detectors will support one of general relativity's key predictions. Michelson was the greatest experimenter of his day. His interferometer was capable of

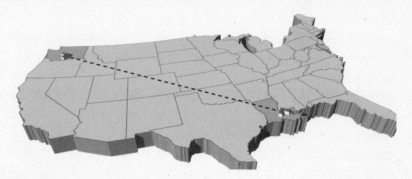

To rule out local disturbances, the LIGO sites are separated by 1,900 miles.

detecting a length change equal to one-twentieth the width of a light wave. In LIGO's early phases, researchers anticipate doing nearly a hundred billion times better. In terms of measuring a change in distance, it is currently science's most sensitive instrument. The very nature of this endeavor seems paradoxical. To detect the very small— shifts in space-time smaller than a subatomic particle—requires an instrument that is very big, 4 kilometers long. To boost the sensitivity to such exquisite levels, researchers have applied an added trick: the laser beams complete the round-trip up and down the arms more than a hundred times, a total of some 800 kilometers (500 miles). This makes the arms act even longer, which increases the chances of detecting a feeble gravity wave.

Working together, the Hanford and Livingston observatories form a single instrument that slashes diagonally across the continental United States. Separated by 1,900 miles (or a hundredth of a second at the speed of a gravity wave) the two locations were among 19 sites proposed in 17 states. The final choice was made through a combination of politics and scientific practicalities. The chosen areas, of course, had to be fairly flat. They also had to be seismically and acoustically quiet, have such amenities as telecommunications and nearby housing readily available, and be set at least 1,500 miles apart to eliminate signals generated by local noises. As with Weber's setup, when the vibrations from a passing train, truck, or other jostling event show up in only one of the detectors and not the other, it can be ignored. Operating in unison, the two facilities have the potential to vastly broaden the search for gravity waves. Bars now in operation can theoretically detect a supernova going off in our galaxy, but the chances of that occurring are slim, once every 30 to 100 years. Two neutron stars colliding somewhere in the Milky Way occurs only once every 100,000 years. That's not good odds for a steady career. LIGO was sold on the idea that it will expand the search to the myriad galaxies beyond our galactic borders, so researchers will have a better chance of seeing gravity wave events on a more regular basis.

The structures in Washington and Louisiana are very similar but more like fraternal twins than identical. The Hanford facility actually houses two interferometers, which operate side by side through the arms. There is the full-length interferometer of 4 kilometers but also one half as long. Midstations situated 2 kilometers down the arms accommodate the end masses for the truncated detector. With different optical configurations, each interferometer could be tuned like a radio, hunting for gravity waves of different frequencies. It's also the means to get three chances at verifying a gravity wave detection with only two sites. Alternatively, one of the Hanford interferometers can at times operate round the clock, always on the lookout for gravity waves, while the other setup is used for tinkering, allowing researchers to learn more about the technology while they steadily upgrade the equipment. The dual system offers great versatility.

The LIGO detectors on each coast are best receptive to frequencies from 100 to 3,000 hertz. On the musical scale that roughly extends from an extremely low A note to a very high F-sharp. Within that broad band observers will be looking for all kinds of signals from the cosmos. Given the audio range of the signal, there might be single cymbal crashes from exploding stars, periodic drumbeats from a swiftly rotating pulsar, an extended glissando—a rapid ride up the scale—from the merger of two black holes, as well as a faint background hiss, the gravitational equivalent of the cosmic microwave background (more on gravity wave events in the chapter entitled "The Music of the Spheres"). Moreover, LIGO will always be comparing its findings with other heavenly data gatherers, such as bar detectors, astronomical telescopes, and neutrino detectors, just in case another spectacular event, such as Supernova 1987A, occurs on its watch.

A true gravity wave washing up on the shores of Earth will affect both detectors simultaneously. But in the process the interferometers face an array of interferences that Weber never had to worry about, which makes this endeavor seem all the more astounding. Take ocean waves, for example. Each LIGO detector is located miles from a sea, but that doesn't prevent the planet's bodies of water from introducing a noise. When waves hit the shores of North America, all over the edges of the continent, they collectively produce a low reverberation roughly every six seconds, a microseismic growl that peaks at 0.16 hertz. That's a low note to best all low notes. In fact, it's so low that the vibration travels through the Earth with ease. "If you just have a mirror sitting there," says LIGO researcher Michael Zucker, "it feels that rumble." So LIGO's mirrors get a tiny push. "It's a headache, a huge headache during Louisiana hurricane season," adds Zucker. LIGO is also affected by "Earth tides," deformations of the Earth caused by the gravitational pull of both the Sun and the Moon. The effect is slight— several millionths of an inch—but still noticeable enough that LIGO has to take it into account. "Tidal actuators" cyclically push and pull on the optical tables to compensate.

More worrisome, perhaps, are thunderstorms within the vast heartland of the United States. It's the one outside source of interference that does have a chance of being felt simultaneously by both widely separated detectors. When a thunderstorm, say in Utah, unleashes a lightning bolt with millions of amps of current, that bolt produces a magnetic field that propagates over the entire country. It's possible that both sites will see it at the same time. Such a coincident signal could be misinterpreted as a gravity wave event. To filter out such noises, magnetic coils have been set up at each site to monitor these magnetic pulses. No CB radios or cell phones are allowed on the sites as well, since they might also interfere.

Two decades ago LIGO started out with a small complement of researchers. Today, it has evolved into a vast enterprise. Its final construction cost of $292 million, plus nearly $80 million more for commissioning and upgrades, made it the single most expensive project ever funded by the National Science Foundation. (The doomed Superconducting Super Collider, which was to cost some $8 billion or more, was mostly funded by the U.S. Department of Energy.) Along with LIGO's in-house staff, working at Caltech, MIT, and the two detector sites, are researchers at other universities who are also funded to actively work on future improvements. Together, they comprise the LIGO Scientific Collaboration. Their job is largely to conquer the array of instrument noises that can potentially hide a gravity wave signal.

Stan Whitcomb has seen this development firsthand. Having previously worked on instrumentation for submillimeter radio astronomy telescopes, he first came to Caltech in 1980 to help put together the 40-meter prototype designed by Drever. Within five years, though, he left for industry, partly due to his discouragement over the slow rate of progress. "In 1980 we had assumed that we'd have large detectors in place within eight years. But we hadn't recognized all the technical problems. We didn't appreciate how difficult it would be to make the sensitivity improvements. We might not have gone forward if we knew ahead of time," he says. "We have six chunks of glass—the four test masses, a beam splitter, and a recycling mirror—all suspended on wires. We cannot touch them, yet we must still be able to position them to within a billionth to one hundred-trillionth of a meter." He returned in 1991 when he saw NSF make its stronger commitment to the project. Advances in technology made him more optimistic as well. "We were blissfully ignorant and lucky that technologies have since advanced to help us—advances in supermirrors, lasers, and vacuum systems," he says. Each piece of the detector is a marvel of engineering.

Mirrors

GariLynn Billingsley is the guardian of the mirrors. Working from Caltech, she monitored each and every step of the production. "Their manufacture was a heroic effort," she says. These pieces of glass pushed the limits of mirror construction, as their specifications were well beyond normal industry standards. Since the end mirrors are parked 4 kilometers away from the laser, they must reflect the light up and down those long corridors extremely accurately. What was required was a mirror surface so smooth that its surface does not vary by more than 30 billionths of an inch, especially in the central 2 inches where the laser beam will be aimed. "Imagine the Earth were that smooth. Then the average mountain wouldn't rise more than an inch," points out Billingsley. The process started with choosing a material that helps reduce one of the major sources of interference in an interferometer: thermal noise. At room temperature the atoms within the mirror are continually vibrating, movements that could easily mask an incoming gravity wave. But if the mirror material acts like a bell (the engineering term is having a "high Q" for quality factor), those jiggles are confined to certain narrow frequencies. Restricting the noise to those specific bands allows other frequency windows to remain open for gravity wave searches, free of interfering noise. It's akin to pushing the furniture in a room off to one side, leaving the remaining space clear for use.

Fused silica was the material of choice. It is a type of glass that can be manufactured exquisitely pure and uniform. This was carried out by both Corning in the United States and Heraeus in Germany. The glass was then sent to both General Optics in California and CSIRO (the Commonwealth Scientific and Industrial Research Organization), a government lab in Australia, for polishing. What resulted were cylindrical disks 10 inches in diameter and 4 inches thick. Each weighs 22 pounds. The final step was applying the reflective coating, a job performed by Research Electro-Optics in Boulder, Colorado, expert in low-loss mirrors. The thin covering is composed of alternating layers of silicon dioxide and tantalum pentoxide. For every million photons of light hitting the mirror, only a few are lost with each bounce off this surface. Twenty-four mirrors were ultimately made. Each interferometer uses four of them as test masses. The rest are spares. In addition, an assortment of secondary optical pieces were polished and coated to the same standards for use in aiming and guiding the laser beam in the interferometer.

The spares are kept in the LIGO optics laboratory, situated in the bowels of the Bridge Building on the south end of the Caltech campus. The disks are stored in what look like aluminum cake covers set on shelves in earthquake-proof cabinets. The lab itself, a windowless room, resembles a hospital operating room. Visitors and technicians wear masks, caps, and booties at all times to keep the environment clean, for just a short breath on a mirror would seriously contaminate its surface. Looking down upon a glass disk is like looking into a pond of pure, perfectly still water. To the naked eye there is nary a bubble or scratch. "I've lifted the glass some 95 times so far and every time I hold my breath until I get it down again," says LIGO optical engineer Steve Elieson. He has good reason to be nervous: each mirror costs $100,000 to manufacture from start to finish.

Suspension and Seismic Isolation

The mirrors are hung from what looks like a gallows. By hanging freely, the mirrors feel no other forces except gravity. It separates them from the rest of the equipment. Mounting them in this way, though, was a nerve-wracking process for the engineers. Each cylindrical mirror is balanced on the slimmest of supports: one steel wire, as thin as dental floss, that is attached to the gallowslike frame. This fine wire is very similar to guitar string, just one-hundredths of an inch thick. Like the silica of the mirrors, this wire has a high Q as well. In this case, heat causes it to vibrate like a stringed instrument at around 340 hertz. The material was chosen to keep both the tone as pure as possible and the string vibrating as long as possible. It's like a having a concert where the single tone of a violin doesn't die down for several minutes after the music is over. In this way a gravity wave with a lower or higher frequency can be better distinguished and not hidden within a cacophony of thermal noise.

Cradled on its wire like a child on a sling swing, the mirror can actually move back and forth without disturbing the measurement. Gravity waves will cause the mirrors to move quickly, anywhere from 100 to 3,000 times each second (100 to 3,000 hertz). (Or looking at it from another perspective—a relativistic one—the space measured *between* the mirrors will jitter.) The movements induced by normal geologic processes, on the other hand, are relatively "slow." They can cause a mirror to oscillate just one time each second. The motion is usually quite tiny, the width of a bacterium. Such extremely low-frequency motions are hard to filter out. But if the mirror should do this, move back and forth over one second (a 1-hertz vibration), observers can simply ignore the motion, essentially subtract it out. It's one of the tricks that enables researchers to zero in on the incredibly tiny space-time movements introduced by a gravity wave. Seismic and tidal movements are so sluggish, compared to a gravity wave signal, that they don't mask the cosmic signal at all.

Nevertheless, LIGO must be isolated from any number of ground motions: a work truck driving by, a toilet flushing in the main build-ing, or a seismic tremor. Such noises might still introduce a vibration that mimics a gravity wave. The first line of defense is the floor itself. It is a slab of concrete, 30 inches thick, which is not coupled to the walls, so any outside vibration—such as a good stiff wind—will not rever-berate into the system. As a second line of defense, all the masses and other optical equipment, housed in the vacuum chambers, have their own seismic isolation system. The optics rest atop a series of four monolithic stainless steel platforms, stacked one on top of the other. Each level is separated by a set of elastic springs, which highly damp any ground motions, not unlike the suspension in a car. It cuts down the seismic noise by a factor of a million.

Vacuum System

The largest portion of the funds for LIGO, about 80 percent, did not go into sophisticated equipment, such as electronics or comput-ers. Rather, it went into the low-tech items of this high-tech enterprise: constructing the pipes, laying the mortar, and building the vacuum pumps. The stainless steel tubes were fabricated by a method regularly used to make oil pipelines. Temporary factories were set up near both LIGO sites to carry out the task. In Louisiana the beam tubes were constructed in a vast warehouse next to a shopping center some 20 miles from the observatory. Over 11 straight months, Chicago Bridge and Iron workers produced 400 separate pipes, each 65 feet long. The stainless steel arrived at the warehouse on immense rolls, like giant-sized rolls of kitchen foil. On an automated conveyor, a roll unwound, with the continuous sheet sent into a machine that coiled it helically and then welded it automatically with a high-frequency pulse of energy. Rolling along at 40 inches per minute, a 65-foot-long section could be completed in about 45 minutes. The finished tube shimmered with stripes of dull red, blue, and cream, resembling a designer barber pole. Each pipe was individually tested for leaks (not one failed) and then taken to its final cleanup. In a separate room a small robotic cart was sent into each tube, which washed and steam rinsed the entire interior. In the end what remained was one part dirt for every million parts of water. "Each detail is like this," says LIGO engineer Cecil Franklin. "And any one error can cause problems down the line." Each tube was finally encased in a plastic bag—what looked like a body bag—for transport to the site, where the separate tubes were finally welded in line to form the long arms. All in all a complete interferometer has 30 miles of welds.

The ends of each arm are actually situated several feet higher off the ground than their starting point at the center station. That's to compensate for the Earth's curvature. The gradual rise keeps the laser beam tubes positively straight as the Earth curves downward. With the Hubble Space Telescope debacle on his mind, where a small measuring error in the factory led to an out-of-focus mirror, Weiss was waking up at night in a sweat during LIGO's construction wondering if the tubes were truly arrow straight. Then he realized they could just look down a completed arm and check. "We got down at the end of one tube in Livingston and asked someone to hold a big searchlight at the other end," recalls Weiss. "We could talk to them because the tube is a good acoustic waveguide. The light was turned on, and when we looked it was about 30 centimeters low. We called out, 'Why don't you put it in the center?' They replied, 'It is in the center.' Within 15 seconds, it dawned on me what was happening: bending of light by the atmosphere, due to both the change in air density from the top of the tube to the bottom as well as temperature differences. The laugh is we were lucky. It was early morning, before the Sun really beat down on the apparatus. If we had waited just three hours, we wouldn't have seen any light at all. I would have assumed we had left something in the tube and demanded to go back in, which would have been expensive."

Air is the biggest obstacle to a laser beam, which is why the arms are evacuated down to a trillionth normal atmospheric pressure. That's to prevent light from scattering off stray gas molecules and introducing a noise. With each detector taking up 300,000 cubic feet of space, LIGO ended up creating the largest artificial vacuum in the world. The atoms that manage to remain in the pipes would fill only a thimble under normal atmospheric pressure. LIGO accomplished this feat by not following conventional wisdom. Instead, it took a risk on a radically new procedure. LIGO researchers arranged for a special steel, an alloy that was cooked for several days to remove excess hydrogen to levels a hundred times less than those of commercial vacuum systems. Ordinarily, hydrogen atoms leak out of steel, which can clog up a vacuum. Designers of particle accelerators deal with this problem by heating their pipes after assembly. This excites the hydrogen molecules enough to coax them out of the metal, where they can be vacuumed away. But such pumping would have been a terribly expensive process for pipes 4 feet wide and miles long. Limiting the hydrogen in the steel from the start made the price of LIGO affordable. Though the pipes were still heated to eliminate the last remaining gases, the number of pumps required to suck away those stray molecules was sharply reduced. "Otherwise, LIGO would have been far too costly to construct," says Whitcomb.

Lasers

As Ron Drever discovered early in his investigations, a gravity wave interferometer requires a laser that is as steady as a rock. Any change in frequency or intensity might be mistaken for a gravity wave effect. Originally, LIGO was designed to use an argon ion laser, a type of laser that emits a brilliant green light. "But having the argon ion laser was like using a radio with vacuum tubes after the development of transistors," says Barry Barish. At the last minute they switched to a solid-state infrared laser, which is extremely stable. Small and powerful, they are called neodymium YAG (for the yttrium-aluminum-garnet crystal that lases). Over a billion trillion cycles of light, its frequency doesn't vary by more than one cycle. Moreover, such a laser shows promise in scaling up its power. Initially, LIGO will use a laser with 10 watts of power, but plans are under way to upgrade to 100 watts, which would appreciably reduce an interference known as "shot noise." As Einstein first noted in his Nobel-Prize-winning discovery, light travels in discrete bundles called photons. When those particles of light hitting a mirror are few, the count is rather noisy. Consider the noise emanating from a slow-dripping bathroom sink. The sound of the individual drops is easily noticeable (and quite annoying) when the flow is small. But the noise smoothes out and gets quieter when the flow increases to a steady stream. Similarly, when the laser power is increased in an interferometer, the relative strength of its shot noise is lessened. Recycling the light—returning it to the interferometer some 100 times— also decreases shot noise. By increasing the laser power in these ways, LIGO will be able to "see" much farther out into the cosmos. It's because the strain—the warp in space-time that LIGO can measure— is directly linked to its laser power. A 10 times improvement in laser power—and the concomitant decrease in shot noise—will allow the detector to measure smaller space-time strains arriving from events far more distant.

Interferometer

Together, these varied pieces of equipment form the interferometer itself, the very heart of LIGO. Nothing gets done without its working. Zucker at MIT, who cut his teeth on the Caltech prototype as a graduate student in the 1980s, now heads up the LIGO task group on interferometer control. "It's the glue that holds the optics in alignment, to make sure that the lasers are resonating in the proper way,"

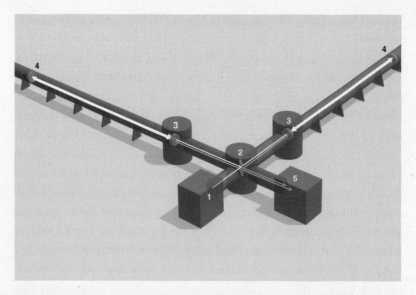

The beam of an infrared laser (1) enters the interferometer and passes through a beam splitter (2) to form two beams. Each beam is directed down an arm and reflected many times between a central mirror (3) and an end mirror (4). Eventually, the beams exit the arms and are recombined. If a gravity wave passes through, altering the length of the arms, a photodetector (5) will detect the change in the resulting light pattern of the recombined beams.

says Zucker. The strategy is to have the photons circulate as long as possible (which improves sensitivity). Based on the current reflectivity of the mirrors, they expect about 130 bounces. In each journey down an arm and back, the waves must march lockstep in synchrony with one another. They must stay "in phase." To do this, exquisite timing is demanded. The software must know exactly what time it is, so every element of the system can stay in synch. Like a human heartbeat skipping a beat, losing track of time would lead to trouble. When the light waves are getting out of step because a mirror has moved, a force is applied to keep the arm lengths equal. In other words, when a gravity wave pushes on a mirror, the controls will be programmed to pull it back into place. The gravity wave signal will actually be hidden in those manipulations. By keeping track of the forces needed to keep the arms firmly fixed, the interferometer will in essence be recording the gravity wave itself. The mechanical movements needed to counter the gravity wave will mirror the strength and frequency of the wave itself.

LIGO has consulted the field of control system engineering to develop its elaborate feedback system. "But we have a pretty unique situation," says Zucker. "Most control system engineers' jaws drop when they hear what we're trying to do." Fine-tuning the position of each mirror is done with tiny magnets. Six in all, each no bigger than an ant, are attached to a mirror—four magnets on the back and two on the sides. Made of rare-earth metals, these magnets are extremely strong for their size. To pitch a mirror forward, its top magnets are pushed while the bottom magnets are pulled. For yaw the magnet on the right side is pushed, the one on the left pulled (or vice versa). To adjust the length of the arm itself, the interferometer can be directed to either push or pull on all four back magnets. In this way the mirrors can conceivably be moved a maximum of 20 millionths of a meter and a minimum of 10^{-18} meter. These are precision movements done nowhere else in science.

Simulations

In particle physics experiments, physicists are often searching for specific discrete events. Particles collide at a fateful moment in an accelerator, creating a burst of energy that instantly transforms into a plethora of new particles. The debris is there to sift through. The long history of particle physics has given physicists a good feel for distinguishing a real event from a fake. In gravitational physics, on the other hand, researchers will be dealing with a continuous stream of unknown data. Where to begin looking? How do you recognize a gravity wave, when one has never before been seen?

LIGO will partly depend on simulations to point the way to the science. A series of programs can mimic all the known sources of interference in the interferometer, such as the seismic, thermal, and shot noises. LIGO researchers are like doctors learning and characterizing every symptom of their patient. When first turned on, LIGO might have a noise level a million times higher than must be attained.

The simulation will help them locate and correct the initial sources of noise, be it a misaligned mirror or a noisy cable line. "There are diagnostic probes all over the instrument," notes Hiro Yamamoto, who directs the simulation efforts. Over time they will come to understand the detector's unique "personality" and maintain a catalog of its typical noises. A gravity wave, it is hoped, will stand out as something different.

Yamamoto first worked on simulations for the Superconducting Super Collider. "It took me some time to adjust my common sense, the way to understand an event in gravity wave physics," he says. But there is some common ground: in both particle physics experiments and gravity wave searches, scientists have to understand the instrument down to the smallest nut and bolt to be able to distinguish the background noise from the signal. "In gravity wave detection, though, we don't know the background as well," points out Yamamoto. "People have thought hard about the possible noises and the perceived noises. But no one knows if you make all those noises go away what is hidden, what will remain. That is the challenge. We're attempting to find a jewel within a dense forest."

Data Analysis

Learning how to handle the immense sets of data that will be flowing out of the gravitational wave observatory is almost as large an enterprise as setting up the instrument itself. Sensors are installed throughout LIGO, all providing data on the detector's condition as well as any signal it might be receiving. These sensors keep track of the laser noise (changes in the light's intensity and frequency, which could mimic a wave), electromagnetic interferences, and any geophysical or terrestrial phenomena that might wiggle the test masses, such as seismic tremors or particularly loud acoustic noises. Data streams in from seismometers, tiltmeters, magnetometers, weather stations, and cosmic-ray shower detectors. "The slamming of a door or flushing of a toilet could conceivably introduce pressures that might affect the masses," notes Albert Lazzarini, although tests show that normal office noise will not likely affect the system. These data are collected in several thousand separate channels simultaneously, tracking an event like a movie. Each frame is a snapshot of an interval in time—the signal plus all the attendant instrument noises. At each site this information continually flows at 6 million bytes per second, 24 hours a day. The hard drive on a home computer would fill up in a matter of minutes. With around 31 million seconds in a year, each interferometer will be gathering some 500 trillion bytes (terabytes) of data annually. The data are shuttled in real time to a bank of computers, where the bytes are filtered, compressed, and ultimately put on tape. LIGO's annual budget for tapes alone is around $100,000. These tapes are transported to Caltech for analysis and storage. The data are kept in a standardized format, a procedure that is also being used at other gravity wave observatories around the world. This will allow each observatory to exchange and compare data, which is vital for confirming a potential signal.

The movements of the mirrors themselves, monitored by the laser light, are recorded on a special channel—the gravity wave channel. It accounts for less than 1 percent of the total data collected, but if a gravity wave comes by, that is the channel where it will be found. Unlike with an optical telescope, no pretty picture of the source is obtained. Visible light waves are quite small compared to their sources, be they gas clouds, stars, or galaxies. Such waves can hit a receiver, say a photographic plate, and produce an image of the celestial object. Gravity waves, on the other hand, are often as large or even larger than their source. A gravity wave with a frequency of 1,000 hertz, for example, spans nearly 200 miles from peak to peak. Such a signal resides in the audio frequency band. You can actually listen to the signal once it is electronically recorded, just as Robert Forward did with his early Malibu detector. Some LIGO workers have analyzed the tapes from the Caltech prototype interferometer in this way. "It sounds like a hiss," notes Lazzarini, leader of the data analysis group. "Actually, a hiss with warbles in it, due to the suspension. It's eerie, in some ways like whale songs."

Computers, not ears, though, will be sifting through LIGO's data. Picking out a definitive signal from a chaotic profusion of bits and bytes is not a totally new endeavor. Though difficult and challenging, it will be similar to the way in which military sonar experts search for the distinctive sound of a submarine amid the many noises of the sea. Essentially, as the data stream comes in, it will be compared to a "template," a theoretical guess at what a gravity wave signal might look like. Take, for example, the case of two neutron stars spiraling into one another. Of course, the exact nature of the gravity waves being emitted from such a system will depend on both the masses of the neutron stars and their orientation as viewed from Earth. So there are many possible wave patterns. To do a proper search, LIGO will have to compare its stream of data against some 20,000 to 30,000 signal patterns continuously throughout the day and night, each pattern representing the waves emitted by differing configurations of stellar mass and at various orientations. Fortunately, computers have now achieved speeds that can handle such a load. One commercial workstation alone can handle from 500 to 1,000 template comparisons in real time. For each interferometer LIGO links a few dozen such stations to form a master machine that handles the search on that detector. "They will be just crunching away all the time," says Lazzarini. If a candidate pops up, it will then be compared with the environmental and instrument channels to see if it was just terrestrial noise.

Both sites will be on the lookout for certain classes of events that don't get repeated—a supernova or gamma-ray burst, for example—by continually comparing for timing, similarity of waveform, and local interferences (to reject spurious coincidences). Events that don't look like false alarms will be identified quickly at both Livingston and Hanford. Rapidly, hopefully in less than a day, the evidence will be reviewed by the science team. If the candidate appears genuine, the astronomical community will be notified to maximize the possibility of catching any electromagnetic radiation emanating from the transient event.

A true challenge, says Lazzarini, will be finding the distinctive and continuous call of a pulsar. This much weaker signal will be buried in a year's worth of data. "But what if the strongest gravitational wave signal," says Lazzarini, "is a belch or burp that arrives sporadically? Then what? You have to assure yourself it wasn't just an amplifier or a bad wire." Those sorts of signals, the unexpected or nonregular, will be the most difficult of all, "but also where the biggest surprises and most profound discoveries may lie," adds Lazzarini.

It has now become routine to compare LIGO to a high-energy physics endeavor in its administration, structure, and complexity. And key managers, like Barish and LIGO's deputy director Gary Sanders, were long immersed in that atmosphere. "One day I was part of a group of 1,000 people, working on one of the Superconducting Super Collider detectors. The next day we were closing shop," says Sanders. Moving over to LIGO, Sanders brought with him a particle physics sensibility, learned from the start of his graduate school days. "You're part of a team, all coordinated, delivering a piece of a larger apparatus," he explains. And that apparatus was a natural evolution from earlier accelerator facilities, which operated quite successfully. The territory was well mapped. Veterans in gravity wave detection, however, have no such assured history to fall back on. No laser interferometer prototype built to date has conducted a continuous observation run longer than several days, and no signals have yet been recorded (although, to be fair, none were expected during those early engineering trials). LIGO scientists are taking a direct leap from laboratory to large facility. LIGO, with its 4-kilometer arms, is a hundredfold jump in size from previous detectors. "It's both an exciting and overpowering change," says Sanders. "Overpowering in that it's a whole new set of tools. But I'm here because the physics is 'classy.' There's almost a romantic attraction, this chance to look at a whole new window on the universe."

Since the field of gravity wave interferometry is still so new, the researchers involved come from diverse backgrounds. Solar physicist Ken Libbrecht, who had moved from Princeton to Caltech after his work with Dicke, eagerly switched from studying oscillations of the Sun to oscillations in space-time. "I looked down the hall here at Caltech and saw all these LIGO people around. I decided to join in," he says. The challenge has attracted young physicists from around the world as well. Walid Majid, a member of LIGO's data analysis group, emigrated with his parents from Afghanistan two decades ago during the Soviet invasion of his homeland. Trained in high-energy physics, he was involved in particle searches at both the Stanford Linear Accelerator and Brookhaven National Laboratory. Yet he readily switched fields. "The standard model in physics was so successful that it was no longer an age of unexpected discoveries," he says. "The experiments were just going after the nitty-gritty details." He wanted to move into a field of physics that still offered surprises. Once LIGO builds up a substantial archive of data, Majid wants to search the tapes for specific events, such as the distinctive periodic signal of a pulsar. He hopes to develop the techniques to recognize and zoom in on the specific frequency of its "cry."

Biplab Bhawal came to LIGO from India. He works on the computer simulations of the detector. First trained in electrical engineering, he went on to obtain a doctorate in quantum field theory but was soon concerned that the exotic topic he was working on wouldn't be verified in his lifetime. Noticing a paper in the *Astrophysical Journal* on the Hulse-Taylor binary pulsar, he decided to pursue gravity waves, even though friends warned him that "searching for gravity waves would be like searching a dark room for a black cat that isn't there." His engineering background came in handy. He wrote a paper on how shot noise might be reduced by manipulating the laser light in a certain way. It got noticed, allowing him to join the new effort. Perhaps he was destined. Biplab means "revolution," a name inspired by a politically minded uncle.

Serap Tilav arrived from Turkey, via the Universities of Delaware and Wisconsin, where she worked on experiments in particle astrophysics. As a postdoc, she set up detectors deep in the ice of the South Pole to catch neutrinos from the Sun. "With neutrino astrophysics I was dealing with the very high part of the energy spectrum, trillions and hundreds of trillions of electron volts. Now with gravity waves I'm absolutely at the opposite end. I will be wonderfully well-rounded in the end covering the energy spectrum from one end to the other," she says with a smile. Sitting at her desk, studying potential signal patterns on her computer screen, she talks of gaining a feel for the instrument. "Theorists often talk about neutron stars coming together. When they merge, they say, you will see this particular signal. In reality it's not like that. When you go where no one has gone before, you really don't know what to expect. So we have to train ourselves on the detector characteristics so well that when we see something different we can say 'What is that?'" She saw this in her neutrino experiments at the South Pole. The data looked completely different than what was anticipated. The researchers had not taken into sufficient account the true nature of the ice. It turns out that in such arctic conditions the ice was in an unexpected state. They had to recalibrate for its new properties, which affected the particle reactions. "Changing to gravity waves, I feel very young again," she says. "It's like being a student all over again."

LIGO is more than an experiment. It is a case study in advancing technologies. New materials and new equipment had to be designed and constructed just for this pioneering observatory. And research is ongoing. Even as the initial detectors were installed, laboratories around the globe have been working on the next-generation instrumentation. Research and development is being handled by the LIGO Scientific Collaboration, which extends far beyond MIT and Caltech. It includes Thorne's colleagues from Moscow, Drever's former cohorts in Glasgow, groups in both Germany and Australia, and scientists from Stanford, Penn State, Syracuse, and the Universities of Colorado, Florida, Michigan, Oregon, and Wisconsin. They are studying new materials, improving the lasers, and testing new vibration isolation methods.

LIGO will be upgraded. LIGO workers are always mindful of improvements in sensitivity. By making LIGO just twice as sensitive, for instance, observers will be able to register events two times farther out. But that also means the total volume of space accessible to LIGO has increased by a factor of eight. Consequently, they will see eight times more extragalactic events. For their advanced detector, LIGO II, they hope to improve their sensitivity—their ability to see additional sources—by a factor of 10 to 15. (The detectors will be progressing from an initial strain of 10^{-21} to 10^{-22} or less.) That translates into 1,000-3,000 times more volume in the universe to examine. "That gives us a factor of 1,000 for finding some extragalactic thing like a coalescing black hole. You go from one event every 10 years, which is pretty painful, to an event every three days, which is very nice. Reasonably small gains are very important," points out Libbrecht. But the overall performance of LIGO depends on lowering all the various sources of noise together, not just one or two. "It's like a limbo dance," suggests Libbrecht. "You have to lower the bar for all of them."

Consequently, LIGO is a work in progress. The improvement of just one detector element can be highly involved. Take the material used for the mirrors, for example. Is fused silica the best substance possible? One alternative is sapphire, which has certain advantages. For one it has good optical transmission. But such a quality is not what drives interest in sapphire. It is its mechanical properties. Sapphire has a "Q" that is 10 to 100 times higher than silica, which means it resonates for a much longer time at a narrow frequency. That's because sapphire is a denser and more rigid material. Such a high Q could conceivably decrease thermal noise in a laser interferometer by 10 times. But no one has ever polished a sapphire object as big as a LIGO mirror. That's an unknown. The success of LIGO is dependent on a multitude of such individual details: the choice of the test mass material, the type of laser, the method of seismic isolation. Before the science comes a plethora of engineering decisions. If sapphire is chosen, LIGO researchers must then learn how to suspend it properly so that its chief benefit—its high Q—is retained. Each decision affects another in a seemingly endless stream.

What keeps LIGO researchers on course amidst these storms of detail? "People take pleasure in solving these technical challenges," answers Peter Saulson, "much the way medieval cathedral builders continued working knowing they might not see the finished church. But if there wasn't a fighting chance to see a gravity wave during my career, I wouldn't be in this field. It's not just Nobel fever. Maybe it was a risky choice when I was just coming out of graduate school, but now it looks like a good decision. The levels of precision we are striving for mark our business; if you do this, you have 'the right stuff.'"

After finishing his doctorate at Princeton in astrophysics in 1981, Saulson read the book *Cosmic Discovery* by Martin Harwit, a book that stresses the idea that key discoveries in astronomy usually arrive when scientists examine the universe with new instruments. With that in mind, Saulson hoped to work with Rai Weiss on the cosmic microwave background. Weiss had no money to support another

Expected Rates of Gravitational Wave Detections

Event and Region of Space Scanned	LIGO I	LIGO II
Supernova (within our galaxy)	1 to 3 per century	
Supernova (60 million light-years, out to Virgo cluster)		2 to 3 per year
Black Hole/Black Hole Merger (300 million light-years)	1 per 1,000 years to 1 per year	
Black Hole/Black Hole Merger (6 billion light-years)		10 per year to 10 per day
Neutron Star/Neutron Star Merger (60 million light-years)	1 per 10,000 years to 10 per century	
Neutron Star/Neutron Star Merger (1.5 billion light-years)		1 per year to 1 per day
Neutron Star/Black Hole Merger (130 million light-years)	1 per 10,000 years to 10 per century	
Neutron Star/Black Hole Merger (3 billion light-years)		1 per year to 10 per day

The black holes visible to LIGO will be in the range of a few dozen solar masses. A black hole has a larger mass then a neutron star, which results in a stronger gravitational wave signal, hence their easier detectability over neutron stars. Future upgrades should increase LIGO's sensitivity by ten to fifteen times. These upgrades include a more powerful laser, better cushioning against seismic vibrations, silica wiring, and pure sapphire mirrors. The merger of two black holes, an event that LIGO I can at best barely register at 300 million light-years, should be readily detectable by LIGO II at 6 billion light-years. The numbers above display large ranges because they depend heavily on the theoretical assumptions being made, parameters that won't be known for sure until the first detections come in.

assistant on that line of work, but he did have funds for gravity wave research. "I remember thinking 'It sounds dangerous. I'll do it,'" says Saulson.

Saulson assisted Weiss on LIGO's first proposals but after eight years at MIT set up his own laboratory at Syracuse University, which has a long history in gravitational research. It's the stereotypical physics lab. A waist-level shelf runs across one end, littered with the usual scientific debris: an assortment of notebooks and manuals, wire, screws, floppy disks, and pieces of aluminum. Saulson added one artistic touch, a large poster on the wall of an old Chinese painting by the sixteenth-century artist T'ang Yin. It shows a mountain scene, with a scholar in a thatched cottage looking off to the horizon. A poem in Chinese characters scrolls down the left side. Loosely translated it says, "When I eat the mushroom of tranquillity, my soul drifts off." It is the one island of order amidst the cluttered surroundings.

Saulson is concerned with a laser interferometer's thermal noise. He is one of the world's experts on this problem, which many consider the field's most challenging obstacle. There in the basement of the physics building on the Syracuse campus, he and his assistants—at the time one graduate student and a postdoc—are measuring the internal friction of materials used for LIGO's mirror suspension system. There are two ways of testing them. One is to impart a "ring" to the materials and see how long they resonate, much like sounding a gong. They are also squeezing the materials, to see how long it takes for each material to return to its original state. The more internal friction a material has, the longer it takes to recover. For the suspension wires they are testing such materials as steel, tungsten, glass, and diamond fibers. The mirror materials include varied glasses, fused silica, and sapphire.

For the first generation of LIGO, steel wire is being used to suspend the mirrors, but glass fibers would decrease noise 10 times over, if they can learn how to handle such material. That's what Saulson's lab is testing. The tests are conducted in a cylindrical vacuum chamber 18 inches wide, the size of an oversized trash can. At the moment the chamber is open. The heavy steel cylinder had been lifted overhead by a ceiling crane. The experiment remains on the lab table right below. Graduate student Andri Gretarsson has just suspended a long thin string of fused silica, about a foot long, and made it ready for "plucking." Once a vibration is set into motion, the ringdown in a vacuum can last for hours, even an entire day. The problem he is confronting, though, is frustrating. How to pluck the fiber? For a metal wire he was able to use an electrostatic device that imparted a gentle push. He's worried that such a procedure won't work for the delicate glass fiber. Sound waves won't work, since the fiber will be suspended in a vacuum. At the moment they can measure vibrations in the fiber as tiny as 100 billionths of a meter. But eventually they will need to do 10,000 times better to see the thermal movements they are after. "Most of the time experimental physicists are worried about their instrumentation. Here we have that rare wonderful moment when we get to worry about the problem itself," says Saulson.

When new optical telescopes come online, such as the twin Keck telescopes in Hawaii or the Hubble Space telescope, there is usually a celebratory "first light" event, the moment when the instrumentation is turned on and the first picture taken. LIGO's initiation was not so dramatic. Because of the complexity of its engineering and optics, LIGO will require a few years for its initial shakedown before it reaches the point when all three interferometers—the two at Hanford, the other in Louisiana—can work in concert with one another 24 hours a day. Then and only then can the search for gravity waves really begin.

LIGO researchers concede that their first detectors may not register a thing. "We're amateurs in a way," says Weiss. "We just hope we've made all the right decisions." Weiss never worried like this with the COBE satellite because COBE was an extension of past measurements. Everyone knew in some way what to expect. Gravity wave astronomy, on the other hand, is virgin territory. Its scientists have built instruments for an effect that has never been detected directly. Failure, fears Weiss, could stall the field for a very long time. For its critics that made LIGO technologically unjustifiable and premature. However, LIGO was built on the belief that solutions could not have been obtained without first building a full-sized facility to carry out the needed preliminary tests. For now the detectors have only a miniscule chance of observing the only source that seems guaranteed: two neutron stars spiraling into one another. But as noted earlier there is a reason LIGO is called an "observatory." Its builders do not intend to carry out a single experiment but to operate the facility for decades to come, much the way the great 200-inch telescope on California's Palomar Mountain has been used and upgraded since 1948. Improvements will be added, the chances for detection increased, as time goes on and development proceeds. The U.S. decision to start construction was also an important signal to other countries to speed up their own plans for gravity wave observatories. If LIGO had not been funded, it may have stalled the construction of similar facilities around the world. With the decision to build LIGO, a momentum was established for setting up a worldwide network.

Variations on a Theme

*P*isa, a vital port in Roman times, rises 13 feet above sea level on the banks of the Arno River. The city was a seagoing power in the twelfth century, when it became a republic after participating in the First Crusade. At that time its influence extended over the entire coast of Tuscany, sparking economic prosperity and artistic splendor. It was during this period of affluence, in 1174, that Pisa's campanile—what became the famous leaning tower—went under construction: six tiers of arches on a base, each precariously perched one upon the other and topped with an airy belfry.

The white marble tower resembles an over-tall wedding cake about to topple, as if its builders were on a drunken engineering binge. The lean started as the third level was being built. The ground began to sink due to a water-bearing layer underground. A cultural icon, the structure is continually surrounded by tourists, their cameras clicking away. It is here, the legend goes, that gravity first came to be understood in a scientific manner. Although the story is likely apocryphal, it is said that Galileo dropped balls of various weights from the top of the campanile to prove his new view of gravity. Until then Aristotle's word was the standard law of physics when it came to falling objects. The Greek sage had declared that the heavier a mass the quicker it falls to the ground. But from tests he conducted, Galileo concluded that wasn't true. He figured out that the duration of a fall is independent of the mass. Neglecting air resistance, a tiny marble will fall just as fast as a heavy bowling ball. In his *Dialogues Concerning Two New Sciences*, Galileo had one of his characters, Sagredo, describe the test: "I . . . who have made the test can assure you that a cannon ball weighing one or two hundred pounds, or even more, will not reach the ground by as much as a [hand] span ahead of a musket ball weighing only half a pound." Galileo begat Newton who begat Einstein. Our modern understanding of gravity began with that one elegant thought.

The leaning tower had been closed off for a while. A series of weights were attached to its base on the north side to counteract the ever-increasing lean. While officials in Pisa worked to save the legendary gravitational test site, others nearby were readying a venture to push gravitational research into the future. A French-Italian collaboration is building a LIGO-like detector, known as VIRGO, on the vast alluvial plain just outside Pisa. Although the LIGO detectors are capable of making a gravity wave discovery on their own, since the two separated sites offer the opportunity to reject local disturbances, the biggest scientific payoff will arise when LIGO also operates as an element in an international network of gravity wave detectors. VIRGO will be part of this network, along with other detectors as well. A British-German team has erected GEO 600, a 600-meter-long interferometer, near Hannover, Germany. Japan, meanwhile, has constructed a detector with 300-meter-long arms. Australia also has plans for a large-scale interferometer.

Together, these facilities will eventually form a worldwide system, akin to the global network of radio telescopes that allow radio astronomers to coordinate their observations. Bar detectors are not excluded. Bars and the newer spheres will enhance the network by looking at specific frequencies of the gravity wave signal passing by. When a scientific committee in the United States recommended against the U.S.

building a spherical bar detector, it led to the impression around the world that "bars are out of date," as one observer put it. But gravity wave researchers are generally agreed that their field needs both types of detectors. "Bars are the only detectors giving us data now," points out Adalberto Giazotto of the INFN in Italy. "It seems crazy to stop this line of research. We need to push it until there is a detection with both kinds of detectors. Until then it's hard to compare the two techniques. We also have to consider the astrophysics. Maybe the interesting physics will best be seen with spherical bars." Experts imagine there will eventually be a range of detectors of various designs, each providing a unique contribution toward helping understand the gravity wave signal.

"The search for gravitational waves is a game requiring long, hard effort with a definite risk of total failure," wrote Kip Thorne in 1980. Two decades later that statement remains as true. Detection depends on them all "listening" and comparing notes. The direction to an abrupt bursting source, such as a binary collision or supernova explosion, cannot be adequately achieved with less than three or four observatories, the way a surveyor needs several points to peg a position. Each site acts like a surveyor's stake in triangulating the location. With LIGO and VIRGO working together, it might be possible to pinpoint a source on the sky to within tens of arcminutes instead of a degree. (A single observatory will still have the opportunity to peg the general direction of a *continuous* gravity wave source, such as a pulsing neutron star, by noting the Doppler shift in the signal's frequency as the Earth moves in its orbit about the Sun.)

The VIRGO project is a collaborative effort that involves some 100 physicists and engineers from Italy and France. Construction of the interferometer began in May 1996. Its arms are a bit shorter than LIGO's, 3 kilometers (just under 2 miles) instead of 4 kilometers. Set on flat farmland of clay and sand, these arms extend from south to north and from east to west. The surrounding farmers raise sugar beets, corn, and sunflowers for oil. To the north are the mountains from which Michelangelo obtained the marble for his sculptures. Before laying down the pipes over those many kilometers, VIRGO engineers had to check for lost mines from World War II. One of the war's most important frontal assaults passed right through this site along the Arno River. The central complex is smaller than LIGO's. Instead of one main building, there are four smaller ones, each constructed out of cement and vinyl siding. From afar it looks like a small industrial park. A lengthy ditch, bordered by earthen dikes, runs by the buildings. This channel serves as an emergency runoff should the nearby Arno overflow. In fact, VIRGO was plagued by frogs during its construction, due to water in its basement.

Giazotto is codirector of the VIRGO project in Italy. His counterpart in France is Alain Brillet. Giazotto's offices are located at an INFN center south of Pisa, in a quiet suburb called San Piero a Grado that is surrounded by a national forest and just a short drive from the VIRGO detector. "Here is where VIRGO was born," he says with pride as he arrives at the laboratory. The name VIRGO refers to the Virgo cluster of galaxies. The aim is to have an instrument that will be able to detect supernova explosions as far out as that noted collection of galaxies, which is situated some 50 million light-years distant. The hope is that by sweeping over such a vast volume of space the instrument will spot at least a few sources a year. Giazotto is tall and trim, with thinning silver-colored hair and the bearing of an aristocrat, but his office is decidedly utilitarian, with its one desk, one cabinet, and one set of file drawers. Right outside this spartan office is a walkway that looks down on the laboratory where VIRGO detector equipment is being readied. Like many in this field, Giazotto came over from particle physics. An experimentalist, he worked with synchrotrons to study the weak nuclear force and the structure of nuclear particles. But he sees no great difficulty in jumping to the study of gravity. To him it is still a particle physics problem. "To discover the graviton," he declares is his goal, the theoretical particle that transmits the force of gravity the way a photon transmits the electromagnetic force. "The only way to see them is to build an observatory. And then there's the enormous bonus of better understanding the universe."

He started thinking about gravitational wave detection in the mid-1970s while he was at CERN working on particle physics experiments. By the next decade he was actively campaigning to get Italy into the business of laser interferometers. But he had a decided point of view. From the start he wanted to build an instrument that could detect much lower frequencies than the other systems in the works, and that

meant focusing on the problem of seismic isolation, the largest imped-
iment to detecting low frequencies. He presented the results of his first
tests in the mid-1980s at an annual conference on gravity then meeting
at the University of Rome. There he met Brillet, a French pioneer in
laser interferometry, who was also interested in building a large system.
Eventually, they teamed up and arranged for support from the physics
communities in both countries. Giazotto firmly believes that detecting
lower frequencies is vital for studying certain sources. Take coalescing
binary neutron stars, for instance. "From 100 hertz and upward, you
have just three seconds of observation time before the stars coalesce,"
he notes. "But if you start at 10 hertz, you can get a thousand seconds of
observing. That's why I want the low frequencies." His dream is for
VIRGO to eventually get down to 4 hertz. In terms of wavelength that
would mean gravity waves that span almost 47,000 miles from peak to
peak (one-fifth the distance from here to the Moon).

To get to 4 hertz, environmental motions of the suspension system
must be reduced by 12 orders of magnitude—down to a trillionth of
their original energy. The trick is to stop the outside vibrations from
flowing down the wires that hold the masses and thereby jiggle them.
To do this, the VIRGO researchers have devised a unique seismic isolation system called the "super attenuator." It has no rubber cushioning,
as in LIGO. Instead, there are six circular rings, stacked one on top of
the other to form a structure three stories tall. In certain ways it resembles the multitiered leaning tower, without the lean. Each of these rings
is a mechanical filter, consisting of six triangular metal blades under
enormous tension, enough to block the noise flowing down the wires
to the test mass suspended at the bottom. The scheme appears to work,
at least on the prototype erected in the INFN laboratory. "We shook the
top with a motor to produce a displacement of about 1 millimeter at 10
hertz," says Giazotto. "At the bottom we couldn't detect any change, at
least to a level of 10^{-10}." Ordinary seismic motions, the kind the instrument would face daily, are actually less than that magnitude. Giazotto is
very conscious of the high probability of failure, at least in these first
endeavors. "We are really working for those who come after us," he says.

For VIRGO's future Giazotto favors a controversial proposal:
building a second detector fairly close by, within 30 miles or so. He

points out that with the LIGO detectors so far apart there is the chance that certain types of waves hitting each one will be "out of phase." But with two detectors close to one another, the waves would be "in phase," rising and falling in concert. By adding them together, one would get a decided boost in signal. Of course, there's the added chance that local disturbances would affect both alike, making it difficult to distinguish a true gravity wave. Giazotto believes that a very good isolation system would take care of that concern.

In some ways, VIRGO already has a mate. A second laser interferometer is in place in Europe, the product of one of the world's earliest programs in laser interferometry that has been centered in Germany for a quarter of a century. The German effort in gravity wave detection originated under the guidance of Heinz Billing at the Max-Planck-Institut für Physik und Astrophysik (Max Planck Institute for Physics and Astrophysics) in Munich. Its initial start came almost by accident. At the time the institute included a special physics division devoted to building computers for scientific calculations, a vital need in the days when commercial computers did not yet offer the power to handle complex scientific equations. But as the computer industry caught up to these needs, the institute's computer designers found they were losing their mission. They gained a new vocation when Joe Weber announced that he had detected gravity waves. "The astrophysicists got very excited by this," recalls Roland Schilling, a former member of the computer-building team. "They said it was so exciting that it would revolutionize all of astrophysics if the claim were true." Theorists on staff at the institute wanted to repeat the experiment, but only the computer development group had the necessary laboratory expertise. Almost overnight its members became detector builders for gravity wave astronomy.

The Munich group soon had a room-temperature bar up and running at its facility. The group coordinated its operation with another bar independently built in Frascati, Italy. Within a few years, though, it became evident that Weber's results could not be confirmed. But a spark had ignited. By then there were two alternatives to improve performance: either construct a supercooled bar or switch to laser interferometry. The Munich researchers, wanting to stay in the business but having neither sufficient expertise nor the necessary infrastructure in cryogenics, opted to go into laser interferometry. They were inspired by the work of both Rai Weiss and Bob Forward. They ordered their first laser in 1974 and were highly optimistic. "I remember very well that we originally had a very short timescale in mind. We thought it would take five or ten years," says Schilling. It all looked so promising on paper. But just like Drever in Scotland, the Germans quickly discovered the many pitfalls of the new approach. Their first mirrors were rather bad; their poor surface quality scattered the light, which greatly reduced the sensitivity. They spent years learning how to deal with laser beam jitters as well as figuring out the best way to mount the test masses. At first the mirrors were just clamped onto aluminum blocks, but the linkage was a major source of vibration. Their first interferometer was a mere 30 centimeters long. Afterward, they went to 3-meter arms. Despite the technological challenges, they remained encouraged, especially when Joseph Taylor came to a meeting in Munich in 1978 and announced his first gravity wave results from the binary pulsar. Here, at least indirectly, was proof that their elusive goal was not imaginary.

Throughout the 1970s and into the 1980s the German laser interferometer team held all the records for sensitivity for such an instrument. The Munich group got a head start by utilizing Weiss's initial interferometer design and then reducing everything to its bare bones. Along the way they made many valuable contributions to the technique, such as learning to suspend the masses on wire slings to reduce seismic interference and thus let the mirrors themselves serve as the test masses. These innovations are now part of the standard instrumentation in every interferometric gravity wave observatory either planned or built, but at the time the group was immersed in the Model T era of the field, working out the basics now taken for granted.

Schilling has presently invested more than 20 years in gravity wave detection. Since those early days, a time when researchers knew every worker in the field by name, the community has grown to encompass hundreds of engineers, technicians, astronomers, and physicists, despite the fact that no bona fide signal has yet been captured. "Even if we detect only the sources we expect," says Schilling, "the scientific gain would be more than worth the effort." But there is another major reason Schilling is sticking around for the hunt. "If you compare what has been experienced in optical, radio, x-ray, and gamma-ray astronomy, there were always sources that you did not think of before. There were always surprises. Why shouldn't that also hold true for the gravitational wave business?" he says. It is a sentiment that has become the mantra for the field, its raison d'être.

A turning point for laser interferometry came in 1982. From his work with interferometer control, Schilling came up with the idea for power recycling. At first he thought it wouldn't offer much benefit when incorporated into their design because it required tremendously good mirrors. But Ron Drever was aware that supermirrors were emerging from military technology and independently discovered the same principle: letting the light stay trapped, so that it continues to bounce between the mirrors. This boost makes it appear that the system is using a more powerful laser, which assists in decreasing a major noise in the system. Until this breakthrough, laser interferometry had been the poor cousin to bars in the gravity wave game. Low laser power, scattered-light problems, innumerable vibrations, and poor mirrors made it appear that the technique would remain a dark horse. But by the mid-1980s its status vastly changed. The introduction of power recycling, along with supermirrors and stable lasers, was a critical development. Now laser interferometers are the top contenders, with the bar groups vying to retain a role in a global network of gravity wave detectors. "From the beginning the advantage of the laser interferometers was that they were broadband detectors," points out Schilling. In other words, they can register a wide band of frequencies, whereas a bar is confined to a narrow frequency range.

In 1983, now as part of the Max-Planck-Institut für Quantenoptik (Max Planck Institute for Quantum Optics) in Garching, a suburb of Munich, the team began to operate a laser interferometer with 30-meter-long arms. After focusing on bettering its sensitivity, the team switched its emphasis to more technical issues, using the instrument as a test bed for new technologies. "Problems that had to be solved if you wanted to operate a big instrument," says Schilling. As in the United States, a much larger facility was very much on their minds. At first they thought they would naturally progress up the orders of magnitude. They had already built prototypes with 30-centimeter, 3-meter, and 30-meter arms. It seemed reasonable to try 300 meters. But, politically, they knew they had to jump to 3 kilometers, for that was the size where they'd have a chance to actually detect something, "even though it was against our common sense," says Schilling. A certain attitude was taking root around the world: let's stop playing with the technology and get some results. At the same time the laser interferometer researchers in Glasgow were pitching the idea for a similar facility to be built in Great Britain. With both countries facing budget crunches in the 1980s, the two groups joined forces in 1989. They named the project GEO, which loosely stood for Gravitational European Observatory. "I said we should call it EGO, for European Gravitational Wave Observatory," notes Schilling with a smile, a suggestion politely ignored.

In the summer of 1989 Herbert Walther, then managing director of the Max-Planck-Institut für Quantenoptik, met Karsten Danzmann at a laser spectroscopy conference and convinced him to take over the gravitational wave detection effort in Garching. Danzmann had never been involved in gravity wave physics, although he did have a habit of changing fields every few years. Trained in gas discharge physics, he worked for 10 years on heavy ion collisions, laser spectroscopy, and the properties of positronium. He also had experience building ultraviolet lasers. But he readily accepted the invitation to try this new line of research. Gravity wave research takes a person who is "stubborn, will take a risk, and is hopelessly optimistic," he says in his perfectly accented English, acquired during a stint on the faculty of Stanford University. There is a very subtle boundary, he adds, between genius and being over the edge. "Let's just say that the percentage of people who are on the other side of that edge is a lot higher in this field than in any other," he says with a laugh. He felt right at home. When Danzmann arrived at Garching, plans for GEO were well under way. But within a year the project fell apart. The financial difficulties faced by Germany in the aftermath of German reunification made funds for the endeavor disappear. By 1991 support was cut off. Great Britain also froze funding, due to budgetary problems.

Meanwhile, Herbert Welling, a professor of physics at the University of Hannover in Germany, convinced his university to broaden the scope of its research efforts by setting up a chair in gravitational wave physics, just as Caltech had done. Danzmann took the post in 1993, setting up an outpost of the Max Planck Institute for Quantum Optics in Hannover and transferring many on the Garching research team to the northern town. Once in Hannover, Danzmann began talking with James Hough, who had taken over Drever's position as head of the Glasgow gravity wave detection team, to resurrect GEO but on a much smaller scale. "We refused to die," says Danzmann. They were able to get 10 million marks, less than a tenth of the proposed cost of the original GEO project. "We were sure we could do it for very little money, if we were willing to take risks and do everything in a very unconventional way," he notes. "We just needed a bit of ingenuity." They had to be ingenious. The new project was going to be a fifth the size of their original plan.

Because of its 600-meter-long arms (about two-fifths of a mile), the instrument was renamed GEO 600. Its goal is to detect strains of 10^{-21}, a thousandfold improvement over current prototypes. Danzmann claims they have an advantage in being small. It will enable them to be highly flexible. Equipment can be changed, almost as fast as a new idea is developed, unlike the larger systems where designs must be chosen and locked into years in advance. This allows them to try out interesting new schemes that may well be adopted by the other sites. GEO 600 will act as a testbed for advanced interferometer design. "If you're poor, you have to be smart to survive," says Danzmann. That means they will have a tiny chance to be the first to detect a signal. Or if signals are first detected with the larger systems, a host of smaller observatories, like the low-cost GEO 600, could be built around the world to enhance the gravity wave network.

GEO 600 is located south of Hannover, about 30 minutes by car. It's an agricultural test site, land owned by the state government and operated by Hannover University for agricultural research. The complex was erected right in the middle of working fields of wheat, barley, apples, pears, raspberries, strawberries, and plums. The arms were conveniently built along existing farm roads. Construction was started in the fall of 1995, with a toast of single-malt whiskey, a nod to the Scottish connection. A few drops were sprinkled on the site. "German beer was applied internally later," says Danzmann. Both Hough and Danzmann, close collaborators for many years, are the principal investigators of GEO 600. Bernard Schutz of the Albert Einstein Institute in Potsdam provides the data analysis expertise.

The size of the instrument was dictated by the width of the land available. It might have been called GEO 573. After 573 meters one arm hits the boundary of the allotted land. To make it an even 600, the collaboration is leasing the last few meters from the farmer next door at a cost of 27 pfennigs per square meter per year. That's about 270 marks annually. They have no plans to make GEO 600 longer, turning it into a LIGO. Two hundred meters farther down the arm is the Leine River. "It's an experiment," stresses Danzmann. "It was not meant to last half a century."

If LIGO were described as an extravagant Broadway show, GEO 600 might be called a high school play. To keep costs down, the German-Scottish collaboration depended heavily on technicians on staff at Hannover as well as student labor. "The entire central building is about as big as LIGO's electronics workshop," says Danzmann with a rueful smile. "Contractors did the heavy work, like pouring the concrete and putting on the roof, the bare bones building. We did everything else." Students designed the air-conditioning system for the clean room. It cost 20,000 marks. A commercial system would have cost a million. They also saved money by going with risky designs, such as their vacuum beam tubes, which extend to the north and east. "It's an unproven design that has never been built before. It was rejected by LIGO as being too risky, even though it costs about a tenth of others. We're using a vacuum tube design normally used for air ducts and air-conditioning systems. It's a thin-walled tube, which is stiffened by giving it a corrugation, all the way along. It's like a long bellows. Otherwise, it would collapse. It only has a wall thickness of 0.8 millimeters. It's a 60-centimeter-wide tube. That way you use very little material, so the material cost alone is low, plus the tube is very lightweight, so handling is very easy. We can use students to carry it around. And the tube supports are a lot less demanding because the weight of the tube is like a wet curtain roughly. Heating of the tube is also easy because the walls are thin. Just a couple hundred amperes of electricity makes it hot," says Danzmann.

GEO 600 gets its enhanced sensitivity with a promising new technique: signal recycling. Signal recycling was an idea first fully worked out by Glasgow physicist Brian Meers. You might think of it as an interferometer acting like a bar, being "tuned" to a particular frequency. Consider a radio. You tune into a certain station with your dial, which locks your radio onto a carrier wave at a set frequency. The radio then ignores this carrier wave and looks at its "sidebands," the frequencies right next door where the music and talk reside. A gravity wave telescope in some ways works similarly. The laser light is a very exact frequency. But if a gravity wave passes by, it moves the mirrors and affects the frequency of the laser light. The signal, in a way, is the "music" placed on either side of the laser light frequency. So the light, as it's being circulated within the interferometer, has these sidebands where the gravity wave signal information resides. In signal recycling it is these sidebands that are stripped off the laser carrier wave and sent back into the interferometer, so that the signal can be built up and amplified.

Despite its shorter arms, GEO 600 has the potential to attain a respectable sensitivity, ideally three or five times less than LIGO over broad bands. And for specific, more narrow frequencies, it may closely approach LIGO and VIRGO. The reason is that GEO 600 incorporates engineering features that the larger systems will be using only later. Along with signal recycling, there is an advanced suspension system. The initial LIGO uses a simple, single-pendulum system with steel wires, which is a proven design. GEO 600, on the other hand, uses a triple-stacked suspension system that employs fused silica wires, a design LIGO intends to adopt in future upgrades. "That's the whole purpose of this instrument," says Danzmann, "to push the limits of what you can do experimentally. Of course, if all goes well, it may actually see a signal, although that is not its main purpose."

Also part of the current worldwide network is Japan's TAMA 300 project. This interferometer is located in Mitaka, 12 miles from Tokyo at the country's National Astronomical Observatory. Unlike the other detectors, the 300-meter-long arms of TAMA are completely underground, housed in long concrete tunnels. There are plans to build a 3-kilometer detector later, which might use mirrors cooled to near absolute zero to reduce thermal noise to a whisper. TAMA 300's shorter arms limit its use as a true observatory, but it is already an important development center and testbed for future interferometer technology.

There was a gap in the layout of the observatories. All the early laser interferometry endeavors were in the northern hemisphere, which greatly restricts the geometry needed to pinpoint the location of a source on the celestial sky. With that in mind, Australian researchers contended that an instrument on their continent would enhance the global network and so actively campaigned to build a detector called AIGO, for Australian International Gravitational Observatory. Almost equidistant from the other instruments, AIGO's location improves both the sensitivity and the resolving power of a gravity wave telescope network. Australia opens up a tremendous volume when it comes to cosmic triangulation. "A southern hemisphere detector is a major component of a global array," says John Sandeman of the Australian National University. "One interferometer is not really a telescope. An observatory really requires at least a group of four interferometers."

Australian researchers have now set up their initial detector on Wallingup Plain, about an hour's drive north of Perth. It's a region well traveled by tourists on their way to Australia's famous Pinnacles. This aborigine land is a sand plain, the perfect medium to absorb seismic vibrations. The interferometer is being built in stages. They have started with a prototype with 80-meter arms to test the technology. Later, as funding allows, the arms will be extended to longer lengths, eventually up to 4 kilometers. This will take its viewing range from the Milky Way out to distant galaxies. Like GEO 600, AIGO scientists will be adventurous in testing advanced materials and designs. "A gravity wave telescope is not like an optical telescope that you can build and leave for years without change," says Sandeman. "It's an ongoing process to bring in new technology." He imagines "digging a hole" into the sensitivity curve of LIGO—in other words, a scheme by which they could go to lower and lower strains, far better than LIGO, but over a limited set of frequencies by using signal recycling. They are also examining the use of artificial sapphire, instead of fused silica, for the mirrors and optics to improve sensitivity.

There is the possibility that when these detectors are turned on worldwide for the first time—the lasers activated and data collected—something will be seen right away. More likely, though, it will be a shakedown cruise, a slow learning experience to understand the instrumentation and its various noises. Most are betting that a signal will not be obtained until the second or third generation of equipment is installed later this decade. But when that signal is snared, it will be a sure-fire candidate for a Nobel Prize, and that lure has the potential to turn the field into a high-stakes competition between the various observatories. Weiss is concerned. "We've seduced people into giving us $300 million for a lark," he says. "The same thing is happening in Europe and now it's going to happen in Japan and Australia. If the scientists do business as usual, acting like little impresarios, we're in trouble. If it degenerates into that, I'll be very upset." He wonders whether there will be a carryover from the field of particle physics (where many gravity wave astronomers started out), a mad race to see something before the other guy does.

To prevent such a heated competition, Weiss would like to set up ground rules for confirming a "first detection." At a meeting of gravity wave researchers in 1998 at the Livingston site, Weiss threw down the gauntlet. "The heart of the matter," he says, "will be detector confidence." For LIGO he would want to have the observation occur in all three LIGO interferometers (the two 4-kilometer instruments and the 2-kilometer system) simultaneously. Moreover, there should be a reasonable delay (the 10 milliseconds of travel time) between Hanford and Livingston, as well as no outstanding environmental interference. All sites should see the same spectra, the same frequency, and the same amplitude. "We're spending a lot of money, so it's crucial to be careful," he stressed. Weiss imagines that, as the field matures and detections are more plentiful, the issue of judging a signal will become less volatile. Until that happens, he wants to avoid the conflicts of the past. To the leaders of the world's current gravity wave projects, he offered the following strategy: "A detection of gravitational waves is to be announced only after a statistically meaningful analysis has been performed of the data of ALL instruments that were observing throughout the world. . . . The data and statistical results are brought to a council composed of representatives from each observing group. The initial publication is submitted in two parts. A paper from the group(s) making the observation and their analysis and a second paper from the council discussing the statistical significance in regards to the worldwide effort, in particular, the probability and confidence of detection in some of the instruments as well as the reasons for non-detection in others."

What if, posed Weiss, one group announces and others disagree? "Then," he concluded, "we'd have gone full circle to Joe," referring to the disagreement that persisted between Weber and the rest of the gravitational wave community over what his bars were picking up. The community would be elated if gravity wave detectors were set vibrating, while gamma-ray detectors and underground neutrino telescopes registered signals as well from a visible supernova in our galaxy. That would be the Cinderella scenario, but more likely it will be a tough call. As veterans of high-energy physics pointed out to Weiss at the Livingston meeting, it's more realistic to assume there will be leaks to the press about a possible detection, forcing some kind of announcement. In the world of the Internet and fast publication, there's a short time constant on reasoned deliberation. Even the best scientists can have differing opinions on what level of confidence they will settle on to say something is a "discovery" or just "evidence of." And circumstances can change those assumptions. As one LIGO researcher noted during the debate, "Standards can be lowered as the money runs out."

LIGO was a project that was close to rejection many times yet survived because of NSF officials who valiantly believed in it and fought to keep it alive. Consequently, there will be great pressure to produce results. "But if LIGO puts out specious results, it'll lose its consensus," notes Gary Sanders. To provide extra confidence in a discovery, Weiss would like groups worldwide to refrain from publishing marginal results, especially if one detector sees an alleged signal and others do not. He wants gravity wave astronomers to act locally and think globally. "I look out on all our young people—who now have gray hair. They've been at it for 20 years. It's time we do some science," he says with great emphasis. "This is more than an experiment. We want to learn something deep about the universe." There are examples of cooperation in other areas of astronomy. Astronomers have already agreed that any signals that appear to be contacts by extraterrestrials, for instance, must be confirmed by more than one observatory before a public announcement is made. And resonant bar groups already exchange data and share in publications.

But can an observatory be expected to wait for a lengthy time, responded Barish, before it publishes? Is that truly practical? Unless the signal is particularly strong, determining whether a gravity wave has passed by will be a long process. It may take months, even years, to eliminate all the possible sources of interference and reach a confident consensus within just one detector group. It could take a single group a full year to pull a signal out of the noise through elaborate computer processing. Must they then wait another year for another group to painstakingly rake through their data to do the same? It might damage the field more by withholding a claim than by prematurely announcing a discovery. It might be foolish to await some golden event if the project were in danger of being shut down.

Perhaps more troublesome will be the wait itself. Barish expects there will be critics who lose confidence as LIGO and the other detectors work out their bugs over the years. Gravity wave astronomy, at least at its start, will likely require patience. "The first instrument is not the final instrument," stresses Barish. But despite his technical concerns, he is heartened by the science. "The fact that people can predict gravity wave sources that are within shouting distance makes me feel incredibly confident," says Barish. "Compared to monopoles, these sources are not just optimistic thinking."

The Music of the Spheres

*T*he Pythagoreans were an ancient brotherhood founded by the Greek philosopher Pythagoras, who is best remembered for his famous theorem concerning right triangles: The square of the hypotenuse of a right triangle is equal to the sum of the squares of the other adjacent sides. In the fifth and sixth centuries B.C. Pythagoreans were devoted to such examples of mathematical beauty and extended this fervor into their contemplation of the cosmos. They intently believed that celestial spheres ascended from Earth to heaven like the rungs of a ladder, which majestically carried the planets around and gave forth harmonic tones that created a wondrous music of the spheres. One version of this system had each planet intone a note higher than the one before it, starting with the Moon and work-ing outward to the fixed stars. The seventeenth-century astronomer Johannes Kepler was a devotee of this idea and even wrote down some celestial tunes, audible to God alone, that he associated with various planets in their orbital journeys. Little did he realize that his musical vision would find substance in an entirely different astronomical arena.

For most of astronomy's history, the universe was studied with one means and one means only: collection of its electromagnetic radiation. For the astronomers who came before Kepler, it was with their eyes alone. Later, lenses and mirrors focused and magnified the visible light. By the middle of the twentieth century, astronomy expanded its arsenal of instruments to collect photons from other regions of the electromagnetic spectrum—radio, infrared, ultraviolet, x rays, and gamma rays. At the same time, particles such as cosmic rays and neutrinos began to be gathered from space. With each new technique, notes Penn State theorist Sam Finn, astronomers found something they didn't expect. Radio astronomers found pulsars, quasars, and massive molecular clouds dotting the celestial landscape. X-ray astronomers were surprised by the power of x-ray binaries, which strongly hinted at the existence of black holes. The lesson to be learned, says Finn wryly, is that "astronomers have no imagination."

Given that understanding, who knows what gravity waves might uncover, for they will be offering a far more radical method of gathering information about the universe. That fact is quite apparent when comparing electromagnetic radiation to gravity waves. Electromagnetic waves are emitted by individual atoms and elementary particles. Gravity waves, on the other hand, are generated by the bulk motions of matter. The frequency of the wave is directly related to the frequency of the massive movement generating the gravity wave. At the moment, for example, the two neutron stars in the Hulse-Taylor binary are emitting gravity waves at a frequency of 10^{-4} hertz as they circle one another every eight hours. As they get closer together over the millennia, though, and their orbital velocities speed up, the gravity wave frequency will increase. LIGO will detect the waves from such systems when they match (coincidentally) the audio-frequency range (20 to 20,000 hertz).

The strongest gravity waves of all are created when matter approaches the speed of light, which would occur in supernovas or when black holes collide. This offers the opportunity to explore the most violent phenomena in the cosmos. It would allow astronomers, for instance, to peer into the very heart of an exploding star. This is because gravity waves pass through matter as if it were not there, unlike most electromagnetic radiation, which can be absorbed or scattered by matter as the radiation proceeds on its journey. Electromagnetic waves travel *through* space-time, while gravity waves are actual jiggles *of* space-time itself. With gravity wave detectors, we will be learning about the cosmos from its space-time vibrations. It will be like adding sound to the silent pictures that have been constructed up until this time. Gravity wave astronomers will be listening to the modern-day version of Pythagoras's music of the spheres.

As Weber and other experimentalists made their first forays into gravity wave detection, theorists were active as well. As the number of detectors grew, specialists in general relativity launched their own effort to figure out what "tunes" the detectors might be receiving. First they had to learn how to manipulate the equations of relativity in such a way that they could even address the problem; then they went on to determine the types of waves that might be emitted by various astrophysical events. At the vanguard of this effort was Caltech's Kip Thorne.

Rai Weiss, admittedly no friend of theorists, makes an exception with Thorne. Introducing him one day at an MIT seminar, he noted that "Thorne is one of the most approachable theorists, a physicist who championed LIGO rather than sitting back and doing idle calculations." Thorne even contributed to the instrumentation. He assisted on the idea to serrate the edges of the baffles inside LIGO's beam tubes, to make sure that any stray laser light was properly dispersed. As a theorist, Thorne wears many hats. He has looked into the origins of classical space and time and explored the physics of a black hole. In the public's eye he is most notorious for his work on "wormholes," hypothetical cosmic tunnels through hyperspace that provide shortcuts to both other reaches of the universe and other times. They sound like science fiction but are founded in genuine solutions to Einstein's field equations. Thorne started studying these weird entities in the mid-

1980s because of a friend's request. Astronomer Carl Sagan, then working on his novel *Contact*, had asked Thorne whether there was a scientifically legitimate way for his characters to dart about the cosmos with ease. Thorne and his students came up with a solution that utilized wormholes. Their result, in a paper entitled "Wormholes, Time Machines and the Weak Energy Condition," was published in the prestigious *Physical Review Letters* in 1988. But Thorne's most massive theoretical endeavor in recent years, in collaboration with an armada of graduate students and postdocs, has been carrying out the theoretical needs for LIGO. His Caltech group has been modeling potential cosmic sources of gravitational radiation and estimating the characteristics of the various waveforms.

Born in 1940, Thorne grew up in Logan, Utah, then a small college town of 16,000. Although his parents were Mormons (their ancestors had moved west with Brigham Young), they didn't fit the group's typical conservative profile. Far more liberal, his father was an eminent soil chemist at Utah State University. His mother, who held a Ph.D. in economics, initiated the women's studies program at Utah State and participated in anti-Vietnam War marches. The oldest of five children, Thorne caught the science bug early: "When I was eight, my mother took me to a lecture on the solar system given by a geology professor at the university. I was immediately fascinated. That was my first introduction to astronomy. Before that I wanted to be a snowplow driver. For a little boy growing up in a town in the mountains, where you have snow banks that are six or eight feet high, snowplow drivers are the most powerful people in the world." By the time he was a teenager, a book by physicist George Gamow, *One, Two, Three, Infinity*, hooked Thorne on relativity. Geometry became a passion. He spent many summer hours working on problems in four dimensions. In high school, Thorne was a "cocky kid," as he puts it. Starting in the ninth grade, he sat in on college classes, including geology, world history, and mathematics. If bored in a high school class, he would just get up and leave. "They presumed I was just going off to the university," says Thorne.

Thorne's rebellion continued at Caltech, which he entered as an undergraduate in 1958. For three summers he worked at the Thiokol Chemical Corporation helping design rocket engines for the Minuteman missile program. When asked in his fourth summer to sign a loyalty oath, a legacy of the McCarthy era, he refused. He lost a prestigious National Science Foundation graduate school fellowship for the same reason (although he later received one when the loyalty oath requirement was finally dropped).

His choice of graduate school in 1962 was almost predetermined. Browsing through the physics journals, Thorne immediately recognized that the most interesting work in general relativity was being done at Princeton University, home of John Wheeler. In their first meeting, Thorne intently listened as Wheeler spoke for two hours outlining the outstanding problems of the time. Thorne chose to immerse himself in black hole physics, although that name wouldn't be used for five more years. Joe Weber was also a presence on campus, as he regularly shuttled between Maryland and Princeton to talk with Wheeler, Dicke, and Dyson about the construction of his first bar detector. Thorne finished his Ph.D. in a speedy three years after writing a dissertation on hypothetical relativistic objects that were long, thin, and cylindrical. To his and everyone's surprise, he found that some of these unusual objects would be stable. Today, it's more than an academic exercise. Theorists wonder if the early universe, in the first fraction of a second of its existence, generated similar bodies, now called cosmic strings.

Thorne returned to Caltech as a postdoc in 1965, just as his skills in general relativity were most needed. Two years earlier quasars had been discovered, and some suspected black holes were involved. But others at Caltech, particularly William Fowler and Fred Hoyle, wondered whether supermassive stars were the source of a quasar's power. Fowler funneled students over to Thorne to help him look into such questions. When Fowler gave up his NSF grant in relativistic astrophysics, because of some new duties he took on, Thorne essentially inherited the stipend, which has allowed him to supervise and support nearly 40 Ph.D. students and another 36 postdocs over the past few decades. As a result, Caltech eventually supplanted Princeton as the new mecca for pursuing general relativity. Thorne became well known on campus for his bohemian flair—shoulder-length hair, full beard, colorful shirts, and sandals.

Always Thorne made sure that his work touched bases with the real world. "There was a great richness of things to be done in bringing relativity in contact with the rest of physics," says Thorne. It was the legacy of Thorne's studies at Princeton, where he not only studied under Wheeler but also regularly dropped by Robert Dicke's group to keep in touch with its experimental work. When Thorne was driving across the country with his family to make his move from Princeton back to Caltech, he recalls dropping by the University of Chicago to consult with the noted astrophysicist Subrahmanyan Chandrasekhar (a 1983 Nobel laureate for whom the Chandra X-ray Observatory is named). They talked about neutron stars and gravitationally collapsed objects, which at the time were still conjectures. "They seemed so far from any observation," says Thorne. "But Chandra expressed a confidence that neutron stars and what came to be known as black holes would, in some moderate number of years, be found. That had a big impact on me. I only wanted to work in areas that had an observational backup."

Pulsars were soon discovered in 1967, giving Thorne added confidence that his work on exotic objects would no longer be purely theoretical. But while neutron stars were gaining credence, black holes were still suspect. Many astronomers still clung to the view that nature would somehow find a way to prevent stellar cores—cores heavier than neutron stars—from collapsing into singularities (the upper limit on neutron stars is believed to be anywhere from $1\frac{1}{2}$ to 3 solar masses). Stars lose mass as they age, and some figured that they would always lose enough to drive them under the black hole limit. Thorne was working in an intellectual climate of great skepticism. "Not unlike the skepticism we've seen in recent years about gravitational wave detection, whether waves are emitted sufficiently strong enough so that we will see them," he says. "The intellectual ambiance of that era was very much one that relativity was a beautiful subject but that it didn't have much to do with the real world. It was just a mathematical subject to be pursued for its own intellectual interest. This attitude pervaded physics from the mid-1930s into the 1970s. I found myself following Wheeler's footsteps as an advocate for these things being truly relevant to the astrophysical universe." Thorne's famous text-book *Gravitation*, published in 1973 with Misner and Wheeler, was to a large extent designed to promote the contact between relativity and the rest of physics. "It was in some sense a propaganda piece, as was my public lecturing. I was trying to convince the community that this was a field that was relevant to the rest of physics," says Thorne. What ulti-mately turned things around were the observations arriving fast and furiously from new arenas of astronomy, especially x-ray astronomy. The "final clincher," according to Thorne, was the close examination of an exceptionally bright x-ray source located in the direction of the Cygnus constellation. Cygnus X-1's powerful x rays come from a dou-ble star system consisting of a giant blue star and a dark invisible com-panion whose measured mass of some 10 to 20 solar masses strongly suggests it is a black hole. The x rays are generated as matter, drawn away from the supergiant star, spirals inward toward the black hole.

Starting three decades ago Thorne carved out a special niche for himself in the general relativity community as he and his students began to examine the stability of neutron stars and black holes. "All of this was done in the context that Weber was working on his experi-ment, and we were trying to understand the potential sources for his bar," says Thorne. "His experiments were very much on my mind." At the time there were suspicions that if a black hole got spun up by accretion to a high speed—by stealing mass from a nearby compan-ion—it would start vibrating and tear itself apart. That was one way to avoid having this ugly awkward creature in the universe. But three of Thorne's students, Richard Price, Bill Press, and Saul Teukolsky, looked at the pulsations of a black hole and proved that if you perturb a black hole the disturbance will quickly dampen as the energy radiates away as a set of gravity waves. In the end a black hole remains and very much intact. Thorne showed that the same would be true for a neu-tron star. Thorne had his students visit the math department to bring back new techniques for analyzing the radiation of gravity waves from stars and other systems. In test after test they and their colleagues at other universities came to the same conclusion: The formation of a black hole was inevitable if enough mass was around. Vibrations alone would not stop it. Even if one black hole were thrown at another black hole—one of the most cataclysmic cosmic events imaginable—what results is simply a bigger black hole that is perfectly stable.

In this way Thorne came to specialize in gravitational wave physics. "I have an aversion to working in areas where other people are working because I prefer to do something that is unique. I don't like to be in the position of worrying that if I don't do something today a competitor will solve the problem tomorrow," he explains. Thorne was making a calculated bet. Most people in black hole research at the time didn't have high expectations that the technology would be good enough to see gravitational radiation anytime in the near future and so didn't care to examine its physics deeply. "But I thought there was a shot at it," he says today, still comfortably attired in a loose cotton shirt but his once-long hair now shorn and graying.

It's a tricky business to determine just how much gravitational radiation might be bathing the Earth daily. It strongly depends on the theoretical models for determining how much gravitational energy might escape an event, such as a supernova or a black hole collision. The models are complex and frequently change as differing solutions go in and out of fashion. Thorne has long maintained a chart of the possibilities, with the uppermost line marking off his "cherished belief" boundary, the strongest waves possible without violating any conventional beliefs about the nature of gravity. The potential sources that his Caltech team and other groups around the world have cataloged so far are many and varied.

Black Hole Collisions

LIGO will be after big game. The sources will weigh at least the mass of our Sun but likely heavier. Their movements will be a sizable fraction of the velocity of light, anywhere from a tenth to nearly full speed. The most exciting find by far would be the collision of two black holes. Such a sighting would finally christen black holes as bona fide denizens of the universe. Up to this point, evidence for their existence has been circumstantial. X-ray telescopes regularly pick up signals from remote orbiting bodies, which astronomers interpret as the high-energy radiation released right before a black hole permanently swallows the matter pulled off a companion star. Yet the black hole itself remains unseen. But if two black holes should be orbiting each other, they would at a certain moment reveal themselves. They would eventually spiral into one another, releasing an unmistakable set of gravitational waves that preserves a record of the fateful collision. It would be a cosmic signature unique to black holes.

Picture two black holes slowly circling one another, like a pair of sumo wrestlers warily checking each other out in the ring. Tens of millions of years earlier these two black holes were simply stars, until they exhausted all their fusible fuel and collapsed to the most compact state imaginable. More than mere indentations in space-time, black holes are fathomless pits. No bits of light or matter can climb out of these deep gravitational abysses. That's why ordinary telescopes can't see them, and theorists can do no more than imagine them. Only a gravity wave telescope has a chance of detecting them.

The sighting would occur at one decisive moment, after the two black holes have been slowly orbiting one another, perhaps over millions of years. During that time the pair would have been emitting a steady stream of very weak gravity waves, a wake that continually spreads outward along the canvas of space-time, like the spiraling pattern of a spinning pinwheel, as the black holes circle about. Gradually losing energy in this fashion, the two black holes relentlessly draw together as the years go by. And the closer they get, the faster they orbit one another.

In the final minute of this fateful dance, the gravity waves being emitted become strong enough to be detectable. Instruments on Earth would register a sort of whine, a series of waves that rapidly rise in pitch, like the sound of an ambulance siren that is swiftly approaching. These black holes should not be thought of as masses, points out Thorne. Rather, they should be envisioned as whirling tornadoes of space-time, which are both dragging space-time around them as they orbit one another. "It's like two tornadoes encased in a third tornado, all coming together," he says. As the twirling black holes are about to meet, spiraling inward faster and faster at speeds close to that of light, the whine turns into a "chirp," a birdlike trill that races up the scales in a matter of seconds. A cymbal-like crash, a mere millisecond in length, heralds the final collision and merger. The two black holes become one. A "ringdown," akin to the diminishing tone of a struck gong, follows as the new entity, a pit in space-time that swirls around like the fearsome tornado in *The Wizard of Oz*, wobbles a bit and then settles down. The masses of the two black holes can be determined from the total duration of their phenomenal coupling: the heavier the holes, the greater their attraction to one another, and the faster the merger.

For many years theorists generally assumed that gravity wave telescopes would register a few of these black hole collisions a year, as soon as they were sensitive enough to detect the signals arriving from as far away as the dense Virgo cluster of galaxies, some 50 million light-years distant. More recent calculations by Simon Portegeis Zwart and Stephen McMillan tentatively suggest that there may be more black hole binaries in the Milky Way and other galaxies than earlier suspected, perhaps a thousand times more. Globular clusters, tightly packed groupings where stars are separated by light-minutes rather than light-years, could act as the incubators. More than a hundred of these dense celestial balls, many containing hundreds of thousands of stars, are scattered above and below the plane of our Milky Way (as in other galaxies). Any single black hole produced in such a cluster would eventually settle toward the cluster's core, where it would likely meet other black holes. Gravitationally attracted to each other, the black holes would pair off and form a binary. Circulating within the cluster, some of these pairs might gain enough speed to escape the cluster altogether. Then undisturbed outside the cluster, the black holes would gradually spiral in and merge. It is the one incontrovertible way that physicists can finally clinch the existence of a black hole, nature's strangest star. The black hole would give itself away by the melody of its gravity wave "song," the distinctive ripples of space-time curvature

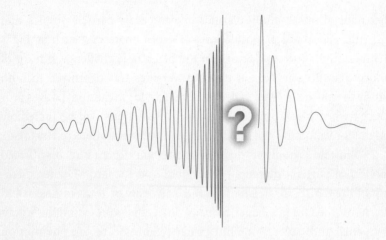

The type of waveform that gravity wave astronomers expect to see when two black holes collide. As the two holes spiral in, the waves increase in frequency. In the end, there is a final "ringdown." No one yet knows what the signal will look like at the moment of collision.

transmitted throughout the heavens. The National Research Council, in its 1999 report *Gravitational Physics*, stated that such a finding "would be the capstone of one of the most remarkable discoveries in the history of science." Listening to these tunes, astronomers might even discern supermassive black holes, each containing the mass of a million or more suns, being constructed in the centers of far-off galaxies, as the holes gobble up and swallow their celestial victims. Gravity wave astronomy will ultimately make black holes seem ordinary, says Thorne.

While the initial whine and chirp phases of the inspiraling are well modeled from numerical relativity, the final crash of the black hole/black hole collision is bringing theorists to the cutting edge of general relativity's challenges. That particular phase of the event is a problem not yet solved in its entirety. Steps toward a solution were carried out in recent years during the Binary Black Hole Grand Challenge, one of a series of problems solvable only by vast computations. Funded by NSF, the other scientific challenges included the formation of galaxies, the synthesis of images from radio astronomy data, the behavior of quark-gluon plasmas in elementary particle physics, and quantum mechanical simulations of exotic materials, such as high-temperature superconductors. In the black hole challenge, a team of more than 30 scientists at eight universities was organized to both study the nature of black holes and predict the gravitational waves produced when black holes collide. Such predictions will be incredibly useful for validating and interpreting the waveforms detected by LIGO.

"Einstein's equations describe gravity via elegant but complicated nonlinear partial differential equations," says Richard Matzner, director of the Center for Relativity at the University of Texas in Austin and leader of the Grand Challenge team. Such equations cannot be solved by pencil and paper alone but rather require brute computation on the world's fastest and most powerful supercomputers. When Einstein's elegant equations are recast in this numerical mode, any one part can involve thousands of terms, which requires special software to handle. Because the solution is so complicated, the team tackled it in steps. First, team members handled the "simple" case of two non-spinning black holes approaching one another head-on, meeting, and then merging. At each stage—from far encounter to near touching—the gravitational landscape changed, an evolution they captured step by step. But as far as nature goes, that's an unrealistic simulation. Like all other stellar objects, black holes spin. And when two black holes are involved, they will also be orbiting one another, which adds more complexity to the numerical simulations. The game in the challenge was to develop the tools needed to first model the process and then optimize the calculations. This optimization depends on both the way the calculations are set up and the power of the computer. In the early 1990s scientists would have needed 100,000 hours of computer processing time (more than 11 years!) to perform an accurate three-dimensional simulation of two black holes spiraling into one another, which made the endeavor wishful thinking. Today, using algorithms optimized for parallel processing machines, that time has been cut to 1,000 hours, a much more doable prospect. But computer memory still needs to be increased a hundredfold before the full solution of a black hole/black hole collision—finding out what happens during the

last few orbits and the ultimate coalescence—can be obtained. Sam
Finn calls it "the final uncharted frontier" in black hole physics.

Neutron Star Collisions

One phenomenon that theorists believe is almost guaranteed will
be the resounding crash that occurs when two neutron stars, paired
together in a binary system, spiral into each other as their orbital
dance decays. In 1963, four years before the first neutron star was even
discovered, physicist Freeman Dyson estimated the gravitational radiation expected from such a neutron star pair. "It would seem worthwhile to maintain a watch for events of this kind, using Weber's
equipment or some suitable modification of it," he wrote at the time.
It was a prescient thought. Gravity wave astronomers suspect such
events may turn out to be the bread and butter of their trade. As noted
earlier, the two compact balls of matter in the famous Hulse-Taylor
binary now emit gravitational waves around 10^{-4} hertz. Only in its last
15 minutes of life, when the two neutron stars have finally drawn quite
close to merge, will the waves sweep from 10 to 1,000 hertz, setting off
gravity wave detectors on Earth. But the Hulse-Taylor binary won't be
colliding for several hundred million years. "So we must reach outside
the galaxy to give our graduate students a thesis," jokes Thorne. As
soon as gravity wave detectors are sensitive enough to see beyond the
Milky Way, they will likely pick up binary neutron star bursts in other
galaxies. The length of the LIGO arms was chosen, in fact, ultimately
to see such events out to a few billion light-years and so obtain a good
population of sources.

How often might this happen? That depends on both the statistics and the latest theoretical models. Theorists take the number of
neutron star binaries known to exist in our own Milky Way galaxy
and then extend that outward to encompass the volume of space that
will be observable to LIGO, eventually a vast region spanning hundreds of millions of light-years. The event rate for LIGO I is rather
low, about 10 per century at best. But with upgrades a LIGO II might
see one a day.

These neutron star binaries will broadcast their own distinctive sets of whines and chirps. More lightweight than black holes, a pair of neutron stars will take longer to merge, so the final recordable signal will last minutes instead of seconds. Gravity wave telescopes would register a sinusoidal wave that sweeps to higher and higher frequencies as the two mountain-sized balls spiral into one another. Five to ten minutes before their lethal meeting, the two neutron stars are about 500 miles apart and orbiting one another about ten times each second, at nearly a tenth the speed of light. In the final moments they are severely stretched by tidal forces and revolving around each other as much as 1,000 times a second, dragging space-time around with them. These waveforms—the "chirps"—will contain within them a wealth of information for those who know how to look, such characteristics as the density and composition of the compact star's nuclear matter. As soon as they touch, the two stars are shredded to pieces, possibly releasing a burst of gamma rays.

What happens afterward? No one knows for sure. An ongoing NASA-sponsored study, the Binary Neutron Star Grand Challenge, may help decide. The remnants might coalesce into a new, more massive neutron star. Or if heavy enough, they might condense to utter invisibility, forging a black hole. Only a gravity wave telescope will be able to reveal the final outcome. But once these signals are detected, they would be a boon to cosmologists, who have long been arguing over the universe's size. Current measurements of distance rely on such yardsticks as the luminosity of stars and the apparent size of galaxies, but astronomers continue to quibble over the interpretation of those standard candles. Until recently, estimates of distance varied by factors of two, which was as vexing to astronomers as if geographers could only estimate the distance between New York and Los Angeles as somewhere between 2,000 and 4,000 miles. But by knowing the amount of gravitational energy emitted by inspiraling neutron star pairs about to collide and comparing these estimates with the strength of the waves when they arrive on Earth, astronomers could calculate how far the waves had to travel to reach our planet. This, in turn, could provide a measuring tape directly out to the galaxies, without the worry of the intervening steps that can plague other methods.

A collision between two neutron stars would also offer insights into the nature of nuclear matter. A neutron star is, in some sense, one big atomic nucleus. Only in this case the nucleus contains 10^{57} neutrons. "Somewhat larger than physicists are normally accustomed to," notes Thorne. So to study this particular form of nuclear material, physicists can't rely on particle accelerators to push the matter to near the speed of light. Fortunately, though, neutron stars in a binary system do it themselves due to their mutual gravitational attraction. Here would be a means of finding out whether some of the mysterious gamma-ray bursts in the sky are the result of two neutron stars colliding, as some suspect. Their distribution and intensity are in the right range. Roughly once a day a burst of gamma rays appears from some random direction on the celestial sky. On average, the flash lasts some 10 seconds. Such bursts were first noticed by U.S. Air Force Vela satellites, launched in the 1960s to monitor nuclear explosions on Earth. While evidence suggests that most bursts originate from far outside our galaxy, their exact origin is not fully known. One of the brightest bursts to date, uncovered by the Italian/Dutch satellite BeppoSAX, is an example. This particular burst was traced to a faint galaxy situated some 10 billion to 12 billion light-years away, nearly to the edge of the visible universe. Over a matter of seconds, it appeared to release several hundred times more energy than a supernova. For that one moment it was as luminous as the rest of the universe. As soon as the burst was spotted in December 1997, an armada of ground-based and space telescopes—optical, radio, x-ray, and infrared—aimed their instruments in that direction and recorded the afterglow of this mighty celestial fireball. It's difficult to explain the tremendous energies; they might involve neutron star or black hole collisions or perhaps the collapse of a massive star to form a black hole. Maybe both. A LIGO detection will help sort it all out.

Supernovas

There could be another type of signal in the gravity wave sky as well, although it would be far less frequent. A solitary tsunami of a wave might hit our shores every once in a while, generated at the very moment a star explodes in our galaxy as a brilliant supernova, its core crumpling up to form a dense neutron star. This collapse triggers a shock wave that blows off the star's outer mantle of gases, which we see as a supernova. Astronomers routinely spot these explosions in other galaxies with the aid of telescopes. For laypersons using only their eyes, it requires a little more patience. The last supernova visible to the naked eye occurred in 1987. The explosion was sighted in the southern sky within the Large Magellanic Cloud, a satellite companion to the Milky Way. The last visible supernova in our galactic neighborhood before that was spotted by Kepler in 1604.

Sighting a supernova with gravity wave detectors is not assured, though. A lot will depend on the explosion's dynamics. If the collapse of the remnant core is perfectly smooth and symmetrical, gravity wave astronomers will not hear even a whimper; gravity waves emitted symmetrically tend to cancel each other out, much the way out-of-phase light waves do. At the same time that one part of the wave is causing space to stretch, another part is causing it to contract; the net result is no change at all. Gravity waves would be emitted only if the collapse is a messy affair, with the newborn neutron star squishing down like a pancake and then stretching out before settling down. As a result, a gravity signal with waves extending hundreds of miles from peak to peak will be sent out. If the core is spinning madly at the end of its life, it could even flatten and be turned into a barlike configuration, spinning end over end like a football. In that case the collapsing core would send out very strong waves, perhaps being seen far beyond the Virgo cluster. An advanced LIGO detector, with more sensitive equipment, might see several a year, which would make it one of LIGO's more dependable sources.

There is some evidence that supernova explosions can be imbalanced. Astronomers have seen individual pulsars speeding through the galaxy at velocities greater than 100 miles per second. It is suspected that the pulsar got shot out by an asymmetric explosion, an extra kick on one side more than the other. Supernovas occur within our galaxy only two or three times every century on average, but their signal strength under these circumstances would be spectacular. It is estimated that the supernova seen in 1987 had a signal 100 times stronger than LIGO is capable of detecting when it first turns on. The unstable neutron star should even "boil" vigorously for the first second of its life. During this boiling, high-temperature nuclear matter (some trillion degrees) rises to the surface, where it cools and is swept back downward. Such boiling could send off a series of 100-hertz waves, strong enough to be seen by a LIGO-type detector out to 100,000 light-years.

Neutron Star "Mountains"

All the while, playing in the background of this gravity wave symphony, could be ongoing rhythms, a steady beat. When a neutron star forms, for instance, it might briefly vibrate and develop a bump on its surface, an inch-high "mountain" that grows and freezes into place. And as the neutron star feverishly whirls around, this deformation, jutting out like a finger, would send out a periodic gravity wave as it continually "scrapes" the space around it. For a neutron star rotating once every thousandth of a second, its equator ends up spinning at about 20 percent of the speed of light. Lumpy neutron stars could serve as gravity wave lighthouses scattered over the heavens, each blinking away until its lump smoothes out. The only interruption might be an occasional gravity wave burp, released whenever the neutron star undergoes a "starquake." This could happen when the outer crust of the neutron star slips at times over its superfluid core. The signal would be extremely weak, too weak to see right away. In this case the interferometer would have to gather the data for weeks and add it up in order to have the signal emerge from the background noise.

More recently, theorists have been excited by a calculation that suggests that a neutron star's dense nuclear matter might actually "slosh around" soon after the star forms, fed by the star's rapid spin. Such oscillations occur in our own oceans, creating circulation patterns. In the case of a neutron star, gravity waves are generated. More interesting, the gravity waves increase the sloshing, which increases the production of gravity waves. What starts out as a quantum jiggle can amplify quickly. How large does the sloshing get? Theorists don't know but suspect other forces, such as friction and magnetic fields, step in at some point to cut it off. Until that happens, though, the newborn neutron star might release a unique gravity wave cry for up to a year as it cools and settles down.

Gravity Wave Background

And beneath the chirps, pops, and beats emanating from the gravity wave sky, there could be an underlying murmur—constant, unvarying, and as delicate as a whisper. This buzz would be the faint reverberation of our universe's creation, its remnant thunder echoing down the passages of time, similar to the residual microwave heat already detected from the Big Bang. But those microwaves started their journey half a million years after the Big Bang—the time when atoms first formed and light could at last travel through the universe, unimpeded by a jumble of particles. If we attempt to look farther back in time, we perceive only a fog.

Primordial gravity waves, on the other hand, would cut right through that fog. They would be fossils from the very instant of creation, tiny jiggles in space pumped up by an explosive burst of expansion that took place a scant 10^{-43} second into the universe's birth. No other signal survives from that era. These relic waves would bring us the closest ever to our origins, perhaps verifying that the universe emerged as a sort of quantum fluctuation out of nothingness. At the same time they might tell us how fast the universe expanded over the eons and whether there is enough matter in the heavens to bring this cosmic marathon to a halt in the far, far future.

It's possible that scientists have already registered the imprint of these primordial waves. In 1992, Martin White of Yale and Lawrence Krauss, now at Case Western Reserve University, first suggested that Big Bang gravity waves may have "ruffled" the cosmic microwave background, which had been mapped so spectacularly by the COBE satellite. COBE discerned tiny fluctuations in the smooth sea of microwaves bathing the universe. Theory suggests that these fluctuations were quantum disturbances that grew and expanded in time. But some of those ripples, say White, Krauss, and others, might be attributed to primordial gravity waves. To separate the quantum from the gravitational requires comparing the COBE data to other measurements of the microwave background. COBE measured fluctuations on a relatively large angular scale. Other instruments, such as balloon-borne detectors and ground-based instruments set up at the South Pole, can measure smaller scales. On those finer scales any contribution from gravity waves should fade away. The background would look a bit less bumpy. Scientists are avidly checking out this possibility.

Perhaps more exciting is the prospect of encountering the unanticipated. The exact form of the gravity wave signal, for one, might disagree with the predictions of general relativity. This may indicate that Einstein's equations have to be amended when dealing with sources that involve a horrifically strong gravitational field. Gravity wave astronomy's findings could usher in new physics of gravity, akin to the way Einstein supplanted Newton. It might even provide clues as to how theorists could forge a "theory of everything" that unites general relativity with quantum mechanics. If not, there is still the chance that strange new celestial creatures could greet us when gravity wave astronomers make their first discoveries. Not until astronomers scanned the heavens with radio telescopes did they discover pulsars and quasars: neutron stars had been contemplated, but not as pulsing radio beacons; quasars were never even fantasized. What else might be skulking about in the darkness of space as yet unseen? Pulsars and quasars may turn out to be commonplace in comparison to the exotic astrophysical events that gravity wave astronomy reveals. Some theorists already wonder whether there might be relics from the early universe, highly energetic "defects" that were generated as the cosmos cooled down over its first second of existence. They include pointlike monopoles, one-dimensional cosmic strings, and what are called domain walls.

Cosmic strings are one of the more interesting defects hypothesized. One might think of them as extremely thin tubes of space-time, skinnier than an atomic particle, in which the energetic conditions of the primeval fireball still prevail. Any strings that survived to this day would be either exceptionally long (spanning the entire width of the universe) or bent back on themselves, creating closed loops that continually lose mass-energy by vibrating at velocities approaching the speed of light. If these potent strings truly exist, astronomers may not be too desirous to observe them close-up. While such a string, anorectically thin, could actually whiz through your body without bumping into one atom, its peculiar gravitational field would wreak havoc nonetheless: if this string sliced through you, your head and feet would proceed to rush toward one another at 10,000 miles per hour. Because of the tremendous tension in a string, it would wiggle around like a rubber band, producing lots of gravity waves. Such gravitational radiation emanating throughout the universe could very well affect the timing of radio pulsars (astronomers are checking). Massive cosmic strings would also be excellent candidates for gravitational lensing.

Meanwhile, x-ray astronomy serves as reconnaissance. Celestial x rays suggest that there's a rich landscape yet to be mined by gravity wave detectors. Take the case of the x rays being emitted by flows of gas close to the center of the active galaxy MCG 6–30–15. This gas is traveling at near the speed of light. At the moment the only explanation for such a high velocity is that the gas is caught in the whirlpool swirling around a massive black hole. But only gravity wave telescopes will be able to tell for sure. They will be able to cut right through the gaseous fog and catch objects falling into the hole directly.

Mounted along the wall of a lengthy hallway directly across from Kip Thorne's office at Caltech is a row of documents, 10 in all, each set in a black frame. Each letter records a bet made by Thorne with some prominent astronomer or physicist, including a few with Stephen Hawking. The challenges are varied, involving either the nature of black holes, the existence of "naked singularities" (a black hole stripped of its event horizon), or some predicted property of the universe at large. One, handwritten on Caltech stationery, concerns the detection of gravity waves. Bruno Bertotti bet a free dinner that gravity waves wouldn't be discovered by midnight May 5, 1988, ten years to the day after the letter was signed. Thorne obviously lost. "Conceded with sad regret," he wrote at the bottom of the letter. But ever the optimist, he had already placed another wager on May 6, 1981, with astrophysicist Jeremiah Ostriker of Princeton that a detection would be imminent:

> Whereas both Jeremiah P. Ostriker and Kip S. Thorne believe that Einstein's equations are valid.
> And both are convinced that these equations predict the existence of gravity waves.
> And both are confident that Nature will provide what physical law predicts.
> And both have faith that scientists can ultimately observe whatever Nature does supply.
> Nevertheless they differ on the likely strengths of natural sources and on the possibility of a near-future and verifiable detection.
> Therefore they agree to wage one case of good red wine.

Ostriker wagered a case of French wine. Thorne went for Californian. The detection had to occur by January 1, 2000. Thorne lost again but is eager to place another bet with any ready taker.

Finale

With test after test currently showing complete agreement with Einstein's general theory of relativity, it's natural to ask why physicists put such effort into designing ever more sophisticated experiments—because, says general relativist Clifford Will, "every test of the theory is potentially a deadly test." Gravity is the force that rules the universe. To understand its workings, to the finest degree, is to understand the very nature of our celestial home. Any deviation, any surprising signal, would surely offer clues to an even deeper understanding of the cosmic design. Consequently, gravity wave astronomers will not be satisfied with earth-based interferometers alone. Interferometers on the ground are limited in the range of frequencies they can detect, which is unfortunate since many interesting events originating from strong astrophysical sources are expected to occur in the very low frequencies, from a millionth of a hertz to 1 hertz. To examine those regions, gravity wave astronomers must venture into space.

In some ways, gravity wave astronomers have already established a presence in space with the many spacecraft that have sped or at this moment are speeding through the solar system on their journeys to various planets. In each case there are two masses: one is the spacecraft; the other is the Earth. From Earth a signal with a very precise frequency is transmitted to the spacecraft, and a transponder aboard the spacecraft sends it back. If a gravity wave passes by and jiggles the Earth, it will shift the frequency of the signal it is sending out ever so slightly. Whenever the gravity wave hits the spacecraft, it will cause a similar frequency change in the spacecraft's returning signal. All in all (as soon as all other sources of noise are subtracted out), what's left is a distinctive set of pulses, the imprint of the radio wave being intermittently altered by the gravity wave. The entire Earth/spacecraft system acts in some way like an interferometer with a single arm hundreds of millions of miles in length.

Planetary scientists have already monitored the communications of the Viking, Voyager, Pioneer 10 and 11, Ulysses, and Galileo spacecraft. The chances were always small that anything would be detected; fluctuations in the solar wind can alter the frequency of a radio beam. But if a particularly big gravity wave had passed by during those missions, the opportunity was there. The gravity waves capable of being found by this method would be extremely long. From peak to peak one full wave could stretch from here to the Sun or even farther. Such a lengthy undulation might have been sent out by two supermassive black holes in close orbit around one another in some far-off galaxy.

One of the biggest tests of this kind was carried out in the spring of 1993. At that time three separate spacecraft were speeding from Earth in different directions, which offered the perfect opportunity to test for gravity waves. NASA's Mars Observer (before it mysteriously malfunctioned) was headed toward the red planet, the Galileo probe was trekking toward Jupiter, and the European Space Agency's Ulysses was journeying toward the Sun. Between March 21 and April 11 of that year, radio signals were simultaneously beamed toward the three spacecraft using a network of radio antennas situated around the globe (NASA's Deep Space Network). Once received, each craft amplified the

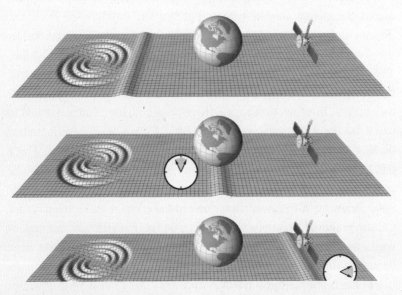

A gravity wave arrives from a cosmic event. One way to detect the wave is to compare how the wave affects the speed of a clock on Earth, then one on a spacecraft situated in the far solar system.

signal and sent it back to Earth. The American and Italian planetary scientists running the test figured that if a gravity wave had passed by, the spacecraft would have gently rocked, like buoys bobbing on the ocean of space-time. That is why the frequency of the radio transmission would have shifted, ever so minutely. A change from just one spacecraft might have been spurious, the result of a local disturbance. But if a frequency shift had been detected in the signals from all three probes around the same time, the evidence would be more convincing. Superaccurate atomic clocks, timepieces capable of discerning a change in frequency of a few parts in a million billion, were used to monitor the potential shifts. Nothing resembling a gravity wave was seen, but that was not too surprising. It was a long shot at best. Spacecraft are still buffeted by the solar wind as they cruise through space, and the Earth's turbulent atmosphere can introduce some radio noise as well.

For the 1993 test the frequencies used to communicate with the spacecraft were between 2.3 and 8.4 gigahertz. But the use of an even higher frequency should help cut down on interference from the plasma in the Earth's ionosphere as well as the solar wind. That will be the case with the Cassini mission, a spacecraft launched in October 1997 and now on its way to study Jupiter and Saturn. "We were hitchhikers before," says John Armstrong of NASA's Jet Propulsion Laboratory (JPL). "But now on Cassini, there is specific hardware on board to carry out these gravity wave experiments." Cassini will use a frequency of 32 gigahertz, four times higher than in previous space tests. There will be three opportunities to search for gravity waves between Cassini and Earth: 40-day periods from December to January in 2001, 2002, and 2003. When such tests were first done in the 1980s with Pioneer 10 and 11, the potential gravity wave strains that the Earth/spacecraft system could measure were around 10^{-13}. Cassini has the capability to do a thousand times better, down to a strain of about 10^{-16}, largely due to its use of the higher communication frequency. That's also 10 to 30 times better than the 1993 experiment. The Cassini experiment, in fact, will be about the best that can be done with this method of testing. To do better will require sending an entire interferometer into space.

In the fall of 1974 that topic came up during a meeting that Weiss convened for his NASA committee looking into gravitational space physics. A group from NASA's research facility in Huntsville, Alabama, made a specific proposal to send a huge laser interferometer into space. Two aluminum trusses, each about a kilometer in length, would be manufactured in space and then put together in the form of a cross. The test masses would be suspended from this structure. It was a time when scientists were eagerly discussing the prospects of sending a variety of instruments into space. After the meeting, over dinner at a Boston seafood restaurant, Weiss asked Peter Bender whether such a gravity wave antenna sounded like a sensible idea. During the ensuing conversation, they began to think of using separate spacecraft rather than a rigid structure, which would allow the mirrors to be many kilometers apart.

Bender was a member of Weiss's committee because of his previous work on a lunar ranging experiment. Trained as an atomic physicist by

Dicke at Princeton in the 1950s, Bender went on to work at the National Bureau of Standards and later with the Joint Institute for Laboratory Astrophysics at the University of Colorado in Boulder. His work on precision distance measurements commenced in 1962, when he began a long-term collaboration with James Faller, another student of Dicke's who had newly arrived at JILA as a postdoc. Faller was eager to convince NASA to deposit a package of reflectors on the surface of the Moon during an unmanned lander mission in order to reflect a laser beam sent from Earth. By measuring the travel time as the beam bounced between the Earth and the Moon, researchers could determine the lunar orbit very accurately, as well as learn much about the lunar interior. Later it came to be seen that the mirrors could be used to carry out tests of general relativity as well, particularly whether the Moon and the Earth differed in their acceleration toward the Sun. By 1965 Faller's idea, with others brought on, turned into the Lunar Ranging Experiment (LURE).

Sheer luck allowed the LURE team to piggyback its project onto the Apollo 11 mission, the first manned landing on the Moon. Worried that the Apollo 11 astronauts would not have time to carry out all of their planned experiments, NASA began looking for projects that didn't require much setup time. The LURE proposal was perfect. All the astronauts had to do was set the reflector package on the lunar surface and adjust the mirrors to the proper angle to reflect a laser beam from Earth. LURE has been a very hardy experiment. Indeed, the system is still working today, the last experiment to continue operating from the Apollo program. The passive mirrors just keep on reflecting, as there is as yet no evidence of damage from dust or micrometeorites. Laser tests with it continue to be carried out from a French observatory near Grasse and from McDonald Observatory in Texas. "It's tested Einstein's Strong Equivalence Principle to a part in a thousand. If you look at tests of relativity, that's one of the major ones that's been done," notes Bender. It shows that gravity accelerates objects equally, regardless of their mass or energy. The Earth and the Moon have been found to accelerate toward the Sun at the very same rate. It's the space-age extension of the leaning tower experiment.

That was Bender's first taste of experimental relativity. His dinner with Weiss in 1974 launched a far more serious involvement. Their conversation about the Huntsville proposal for a massive space-based interferometer soon expanded in subsequent weeks to include both Faller and Ron Drever. "We had recognized that the ends of the interferometer should be separated as far as possible—that you didn't need a fixed structure," explains Bender. "We finally got up our nerve to talk about a thousand kilometers or so, without realizing that others had mentioned such a distance with separate spacecraft considerably earlier." In the early 1970s a number of scientists, including Forward and his colleagues at Hughes, theorists Press and Thorne, as well as Braginsky in the Soviet Union, remarked in print on the possibility of taking interferometers into space as separated spacecraft, not long after Weiss had put his first laser interferometer design down on paper. "Then Ron said, 'Why stop there?' You just gain by making it longer. We ultimately concluded that a distance of a million kilometers was reasonable to think about it," recalls Bender. Here was the kernel of an idea that would evolve and gestate for more than 20 years.

Two decades ago the idea of launching a gravity wave detector into space seemed almost fanciful. It was partly hampered by the impression that laser interferometer technology first needed to be developed on the ground. At the time, stable lasers had relatively short lifetimes and low power. With no improvements, the arms of a space interferometer spanning some 1 million kilometers would have had to match within 10 meters. Such fine-tuning would have required frequent adjustments to the spacecrafts' positions, interruptions that would make measurements far more difficult. But once the idea was on the table, interested researchers began thinking of ways around these obstacles. Faller, who gave the first public talk about the concept in 1981, came up with a way to subtract out the laser noise. Bender, using his expertise in celestial mechanics from the lunar ranging experiment, worked out the best orbit—a heliocentric path about the Sun.

Faller and Bender kept the idea alive, aided by modest grants from both NASA and the National Bureau of Standards (now the National Institute of Standards and Technology). This money enabled R. Tucker Stebbins, who had been involved in experimental relativity since his undergraduate days, to join Bender and Faller to put a more sustained effort into the proposal. They even gave their proposed space detector a name: LAGOS, for Laser Gravitational-wave Observatory in Space. By 1989 LAGOS received high marks from a NASA committee looking into possible space ventures for astronomy after the completion of its Great Observatories program. But when NASA's advanced research funding got cut a few years later, interest turned tepid. "There was even a joking comment about putting us out of our misery," says Stebbins.

The idea revived, however, when a number of American space interferometer veterans joined forces with a larger group of European gravity wave specialists, including GEO 600's Karsten Danzmann, Jim Hough, and Bernard Schutz. Their formal proposal for a mission, newly dubbed LISA for Laser Interferometer Space Antenna, was presented to the European Space Agency (ESA) in 1993. The project was eventually accepted as an ESA "cornerstone mission," which will fly once the funds are available. The total cost is expected to be roughly $500 million. The ESA decision, though, was made with the expectation that NASA would eventually come on board as an equal collaborator. NASA is now taking a hard look at the proposal. LISA project offices have been set up at both JPL and the Goddard Space Flight Center in Maryland, while LISA science teams have been organized on both sides of the Atlantic. LISA is a strong candidate for NASA's 2005–2010 time frame, a period when the ground-based observatories might be detecting their first gravity waves. If approved, it would be part of NASA's mission to explore the structure and evolution of the universe, similar to past agendas that supported such astronomical observatories as the Hubble Space Telescope and the Chandra X-ray Observatory.

LISA was not the sole contender for consideration by NASA and ESA authorities. For a while another idea was circulating within the agencies, a proposal championed by Ron Hellings of NASA's Jet Propulsion Laboratory. First called LINE, then SAGITTARIUS, and then OMEGA, this plan proposed to put the laser interferometer system in orbit around the Earth rather than the Sun, which makes some aspects of the project easier, such as the launch and telecommunications. In his fifties, Hellings was eager to get something up quickly before he retired. OMEGA called for six spacecraft orbiting the Earth, using existing space platforms rather than special construction. It was to be faster and cheaper, adopting NASA's latest philosophy. The downside was that a geocentric orbit placed the spacecraft in a more severe space environment. There would be added thermal, geomagnetic, and gravitational forces near the Earth, which would buffet the satellites around. LISA supporters prefer a heliocentric orbit, which allows the spacecraft to maintain a fixed attitude to the Sun, simplifying the forces on the masses. For that and other reasons, many working on OMEGA eventually joined the LISA effort.

Before ESA stepped in, the LISA project had partly been a labor of love, with various participants working on it in their spare time and using their own discretionary funds to finance the initial studies. But once ESA as well as JPL provided seed money, "many nooks and crannies of the design were looked at for the first time," says Stebbins. The current design calls for three spacecraft to fly in a triangular formation, with the center of the triangle tracing Earth's orbit. The entire system will perpetually follow the Earth like a faithful companion at a distance of some 50 million kilometers (about 30 million miles). Being so far out in space, there will be no seismic disturbances. The test masses, polished platinum-gold cubes 1½ inches wide, will be free-falling along a space-time pathway carved out by the Sun. The orbits were chosen so that each spacecraft will have solar illumination from a constant direction, which will help maintain a stable thermal environment. Once properly inserted into orbit, the three spacecraft should fly in formation with little adjustment, with just a very weak and steady thrust to counter solar radiation pressure, which would otherwise push the spacecraft like wind powering a sailboat.

The three spacecraft will be separated by five million kilometers (3 million miles). Each will be carrying two lasers and two test masses arranged in the form of a Y, so that each spacecraft can be aimed at the other two. This will allow laser light to be continually transmitted and received along each arm of the triangle. Because of the long distances involved, LISA requires a different approach to interferometry. As the laser light travels the long distance from spacecraft to spacecraft, the

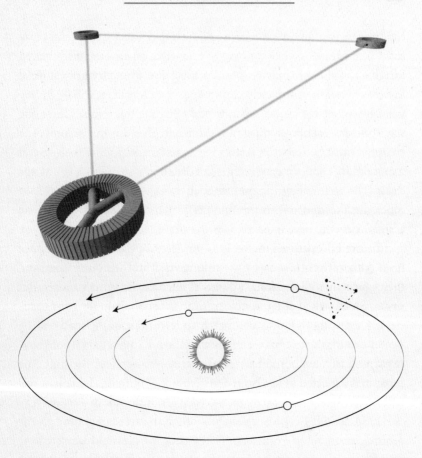

The three LISA spacecraft in formation. Set up in this triangular pattern, each side 3 million miles long, LISA will follow the Earth in its orbit around the Sun.

beam will get wider and wider, eventually spreading out some dozen miles. The original half watt of power will diminish to less than a billionth of a watt at the point of arrival. So the signal cannot simply be reflected back. Rather, the beam must first be amplified—the signal boosted—with the onboard laser before being sent back along the arm. If the beam were simply reflected, only a few photons per second would make it back, which would make measurement impossible. Such an amplification scheme is already used in tracking spacecraft but with radio waves instead of laser beams.

Other engineering needs for LISA require sensitivities and specifications not yet fully developed or flight tested. Most important, there must be a means for maintaining a "drag-free" environment so that all forces upon the test masses (besides solar and planetary gravitational forces) are nearly completely eliminated. Each test mass will be in free fall, isolated as if it were a separate body floating freely in space, so the walls housing each mass must never touch it. Over any one second each chamber must not move in relation to its test mass by more than several nanometers. That's an extremely small distance, a span just a few atoms thick. The technology to conduct such maneuvers has been used on other satellites but not to the fine performance needed by LISA. The adjustments will have to be made with microthrusters, which might use such means as energized metal ions for the propellant. "Essentially we need the most gentle rocket you can imagine," says Stebbins. Moreover, they have to worry about such things as the occasional micrometeoroid strike, which would add a sizable push to the spacecraft. If a dust particle, just a few thousandths of an inch wide, hit one of the spacecraft, it would disturb the test mass enough that the microthrusters would have to turn on to maximum thrust temporarily to correct for the shift. The mission is expected to last from 3 to 10 years. If nothing else fails, it will end when the microthrusters release their last little puff of atoms.

LISA would be a complement rather than a competitor to the ground-based interferometers. That's because it would be receiving very low-frequency gravity waves, from a few millionths of a hertz to 1 hertz, well below the band accessible by LIGO and VIRGO. LISA will be detecting the huge long swells in the ocean of space-time, while LIGO and its related kin observe the finer ripples. Each type of wave is generated by either a different astronomical source or a different moment of an event. LIGO and VIRGO, for example, are best tuned to see black hole/black hole binaries when each hole weighs up to a few dozen solar masses. LISA, on the other hand, will be able to spot black hole systems in the range of 100 to 100 million solar masses. Neutron star binaries, on the other hand, will be visible to both but at different times. LISA will see them as two stars orbiting one another, thousands of years before they collide. LIGO will observe them right before the collision, as the stars rapidly spiral inward and the gravity waves sweep to higher and higher frequencies. Thus, instruments are needed both in space and on the ground to cover the entire spectrum of waves likely to journey through the heavens. A typical wave for LISA, though, will be rather lumbering, taking perhaps 1,000 seconds or more for just one wave to pass by from peak to peak. This is the type of astronomy for those with a patient temperament. Consequently, LISA will ultimately gather fewer data than detectors on the ground. LISA's data rate will be below 1,000 bits a second, versus 6 million bits a second in LIGO. One CD-ROM will be able to store the data from the entire mission.

There is one reason—and one reason only—that supporters have stuck with the idea of a space interferometer through thick and thin financial times. "LISA's greatest strength is its science," says Stebbins. If the technology works, LISA observers are guaranteed to see something. They might even beat the earth-based interferometers in detecting the first authentic gravity wave. From the ground there are many uncertainties about an interferometer's ability to see the sources and how many events will ultimately be observable. Ground-based instruments will be working on the margins of detectability. LIGO should see neutron star binaries coalescing, but the event rate is iffy. While supernovas are assured to go off, the strength of the resulting wave is not well known. LISA, on the other hand, will be overwhelmed by sources, inundated by a fairly noisy background. Galactic binaries, such as the myriad number of white dwarf stars orbiting one another throughout our galaxy, will be broadcasting an unremitting cacophony of waves discernible from space. These sources are considered so surefire that "if LISA would not detect the gravitational waves from known binaries with the intensity . . . predicted by General Relativity," reported a LISA study team, "it will shake the very foundations of gravitational physics."

The expectation is that, once these binary signals from our galaxy are examined and understood, they can be subtracted from the LISA data. "What you're left with is a record that should have the extragalactic information in it," says Bender. That would be the signals from massive black holes in faraway galaxies, poised for collision. LISA would be the instrument of choice for studying such supermassive black holes in distant galaxies. These would be gravity wave astronomy's most powerful sources. Their examination is one of LISA's prime objectives. One of the best routes for seeing the signals from supermassive black holes, gargantuan objects containing the mass of millions of suns, will be catching two galaxies in the act of merging. During this process, the black holes in each galactic core would eventually coalesce to form an even bigger hole, resulting in a tremendous burst of gravitational radiation. LISA should be capable of seeing the entire last year of the inspiraling holes before their fateful collision. Douglas Richstone, a longtime black hole hunter, calls this "the brightest object in the sky that has not yet been seen." LISA has the potential to detect 10 of these events each year. It would be sensitive enough to register waves arriving from as far out as 9 billion light-years, across nearly the entire visible universe.

The chances are quite high that LISA will see something, for evidence has been emerging in recent years that *all* galaxies with a spheroidal bulge, such as elliptical galaxies and most spiraling galaxies, harbor a supermassive black hole in their centers. The larger the bulge, the larger the black hole. Our own Milky Way galaxy harbors a black hole of some 2 million solar masses smack dab in its center. Evidence is mounting that these holes are "quasar fossils," the engines that once allowed each galaxy when it was young to blaze away with the brilliance of up to a trillion suns. It suggests that the formation of a galaxy and the growth of its central black hole are intimately linked in some as yet undetermined way. Perhaps a modest black hole forms first, serving as the gravitational "seed" around which a galaxy forms. Over time such a black hole would consume much of the young galaxy's rich supply of stars and gas, bulking itself up. The hole becomes supermassive. Or perhaps the early universe generated a bevy of smaller black holes, each in an individual galactic building block. These separate pieces could then have eventually merged to form a full-blown galaxy, while the black holes coalesced to form a gargantuan black hole in its center.

In either case, gigantic black holes appear to be the natural result of galactic evolution and serve to power the fireworks display that announces each galaxy's birth. This happens because the massive black hole gravitationally attracts any matter lurking near it and never lets it go. But before this material is permanently captured, it gathers into a swirling accretion disk that surrounds the black hole and radiates intensely. At the same time the hole also spins, like a giant electromagnetic dynamo, producing two jets of subatomic particles that shoot away in opposite directions from each pole at near the speed of light. All of this activity occurs as long as there is enough fuel nearby—stars, dust, and gas—to feed the dark monster in the middle.

Today, quiescent holes can be reignited when galaxies collide. And there are hints in the celestial sky that mergers are in progress. The radio galaxy 3C75, for example, has a set of curving radio jets that resemble a giant water sprinkler at work. These jets appear to be emanating from two nuclei, each possibly an immense black hole. The long spouts get twisted as the two black holes orbit one another. These particular black holes won't be merging for many, many years. But when they at last approach one another, they will emit a distinctive gravity wave signal. The frequency will start low and sweep to higher and higher frequencies as the year progresses and the holes get closer. It will be gravity wave astronomy's ultimate payoff: conclusive proof that black holes are indeed the engines of a galaxy's central activity. LISA could serve as an early warning system. If it can precisely pinpoint the location of an active inspiraling, optical, x-ray, and gamma-ray detectors could be trained on that region to record the final collision.

Almost as interesting will be watching smaller objects—neutron stars, smaller black holes, white dwarfs, and ordinary stars—falling into a supermassive black hole at a galaxy's center. Since stars are so plentiful, this could occur fairly frequently. The final orbits of the doomed stars would be intricate and complex. In some cases, the orbital decay could go on for 70 to 100 years, allowing LISA astronomers to watch them for years. Each object would serve as an exquisite probe for mapping the space-time geometry, the gravitational twists and turns, around a supermassive black hole.

All the while LISA will be listening to that cacophony of gravity wave signals arriving from the host of binary star systems in our own Milky Way galaxy: the waves continually emitted as neutron stars orbit neutron stars, black holes circle black holes, white dwarfs pair up with other white dwarfs, and all the possible combinations in between. Many of these systems cannot be seen with regular telescopes, so gravity wave detectors will at last offer the means of taking a reliable census of these binaries. Astronomers will hit the jackpot if, in one of those systems, a white dwarf merges with another white dwarf, generating a spectacular supernova. By one estimate, LISA has a 2 percent chance of seeing such an event over its lifetime.

What many scientists look forward to are the waves arriving from events even more bizarre. It is a space-based interferometer that will offer astronomers a particularly good opportunity to look back to the universe's origins, farther than with any other astronomical means. Currently, the cosmic microwave background says something about the universe's condition nearly half a million years after the Big Bang. That's when the primordial fog lifted and the universe became transparent. By then the universe had cooled down enough for neutral atoms to form, at last allowing radiation to travel unimpeded. These light waves, mostly in the optical and infrared regions of the spectrum by that point, were gradually stretched with the universe's expansion until today they are detected as a vast sea of microwaves. Before that decisive moment in our universe's history, the primordial fireball was a murky soup—a jumble of protons, simple nuclei, electrons, neutrinos, and photons of electromagnetic radiation, all intermixed. Even if astronomers were someday able to peer back to this epoch, they wouldn't see much, for the cosmic plasma was quite opaque, just as the Sun's hot outer layers prevent us from gazing into its nuclear core. This so-called fireball would be an impenetrable barrier to our view.

But a very sensitive space-based interferometer (if not LISA then its successor) could potentially look back as far as the first hundred trillionth of a second after the Big Bang. Unlike electromagnetic waves, gravity waves can cut right through the foggy layers of the primordial fireball. Moreover, there could be other gravity wave events at special moments in the universe's birth, when the universe experienced an abrupt change in its environment—a sudden shift between two different states, like liquid water turning into something far different, ice. As the universe expanded, it too cooled and likely underwent some kind of "phase transitions." Additional gravity waves might have been emitted in the process. The rapid transition could have given rise to odd defects, areas that retain features of the earlier higher-energy state. That would be the origin of cosmic string. Such strings would be continually oscillating, wiggling, linking up, creating loops. And as they moved, they would generate tremendous wakes, wakes that generate a new supply of gravity waves. Gradually losing energy as they wiggle, they'd finally rocket away in their death throes, disappearing with a final blast of gravitational radiation.

It was surprising, seemingly minor, problems that caused intermittent delays in LIGO's start-up. At Hanford a few weeks had to be spent replacing a glue that failed, the glue that attached the tiny maneuvering magnets to the mirrors. Will LIGO always be at the mercy of such minute details? Fred Raab is not discouraged. "It's like asking the Wright Brothers why they couldn't go up for more than a few minutes at first," he responds.

While the magnets are being reglued at Hanford, Weiss comes out for a visit to measure any residues remaining after the arms had undergone their "bake-out," which cleared the steel tubes of their last remaining gases. An electric current had been sent through the beam tubes as if they were long wires. For about a month the arms were heated in this way to 300° F. The total electric bill was over $60,000. Weiss sets up his temporary office in a small portable trailer, parked right outside Port 5, an access door about halfway down the northern arm. He opens a series of valves to link his gas detector to the tube's interior. He patiently sits at a computer in the cramped quarters keenly watching the screen as a graph gradually displays the level of residual gases left in the tube. The first results are encouraging. "Isn't that pretty," he says, as the program draws its lines. "I don't see a lot from the tube." It was too good to be true, though. A half hour later he notices a subtle change in a curve on his computer screen. It's the distinct signature of a leak. Thankfully, a test the next day confirms the leak was occurring at a port rather than a failed weld, a far more difficult repair. A port leak can easily be fixed by tightening a bolt or replacing a gasket.

Weiss gets a bit nostalgic sitting in the trailer. Nearly three decades have passed since he first sat down and conceived of LIGO's basic design. Now it finally stretches out before him as miles of steel and concrete. "I was here in 1997," he recalls, "when the first beam tubes were being installed. The meadowlarks and magpies would gather right outside. We'd also see swallows fly straight down the tubes, riding on the thermals." After a pause he continues. "A lot of heartache went into this, but it was all worthwhile."

With his settlement from Caltech over his dismissal from LIGO, Drever chose to develop an independent research program for gravity wave detection. He set up his own 40-meter interferometer for tests. He seems more at ease these days, working once again in his own laboratory. "I like to try out slightly crazy ideas," he admits. "LIGO is regimented, which is needed to get it built. But my feelings about that are both good and bad. It only allows for ideas that are guaranteed to work, so its sensitivity is marginal. Whether it will see anything is a toss-up, but the potential is there." His aim is to make the breakthroughs that will guarantee success. For the moment he is working on ideas that will make an interferometer quieter, so that it can be pushed to lower and lower frequencies. "My new lab is for exploring."

His work space is located in what is known on campus as the Synchrotron Laboratory, because a synchrotron—a type of particle accelerator—was once located in the vast cavernous room. Here Drever has set up his interferometer along two sides of the hall. One arm runs the length of the building. The other arm goes through the width of the building and out the other end, where it continues in a tunnel beneath a road. With this detector he has been trying out a new means of suspending the test masses—magnetic levitation. His hope is that by doing away with the suspension wires he can eliminate a major source of noise. Of course, he has to worry about new sources of noise, such as magnetic field disturbances, but he is hopeful. He and an assistant have a prototype working; a small cube hovers without visible support above an optical table, as if by magic. So delicate is the cube's balance that walking nearby causes a tilt in the floor that makes the mass move forward.

By pushing to lower frequencies, Drever envisions his interferometer being used for more than gravity wave searches. "We might see some interesting geophysics as well," he says, such as motions within the center of the Earth. Oscillations of the solid core, surrounded by more fluid layers, should generate gravity gradients that theoretically could be picked up at very low frequencies.

"The people who do many wrong experiments are the best," says Drever. "You try out more things. That's how you make discoveries, but you have to be quick." There is certainly a quickness to Drever's step, as well as in his speech and in his hands. He's always buzzing around, always thinking, always animated. On this particular afternoon he is most excited by an experiment he had conducted the day before, which caused him to stay in the lab past midnight. He took an old phonograph and used its needle and amplifier to make a crude measurement of the thermal noise in a new material he had just gotten a hold of from another Caltech lab, a material that he says could be a breakthrough for LIGO as a test mass.

The history of the hall where Drever now works goes back much farther than the synchrotron. The immense space, windowless and insulated from the heat of the Sun, was actually constructed in the 1930s to polish the massive 200-inch mirror of the famous Hale telescope, which has been dutifully and majestically surveying the heavens for more than half a century atop Palomar mountain northeast of San Diego. When first conceived by astronomer George Hale, though, the telescope's success was hardly assured. As at LIGO's conception, much of the technology was not yet developed to guarantee that its mission could be achieved. Many astronomers at the time were highly skeptical that such a large disk of glass—twice the diameter of Mount Wilson's 100-inch telescope that discovered galaxies and the expanding universe—could be poured or properly mounted and maneuvered. The Bureau of Standards had judged that a telescope larger than 100 inches was a technical impossibility. Polishing its thick slab of Pyrex glass has been described as the Apollo project of the Great Depression era. Just several feet from Drever's detector once stood a gigantic turntable upon which the disk—then the world's largest monolithic piece of glass—sat. From 1936 to 1947 (with World War II imposing an interruption), the turntable rotated while a polishing tool pressed against the disk. Slowly and inexorably the surface was smoothed to exquisite precision while visitors watched from a glass-walled public gallery perched high above the floor. Truckloads of abrasive and jeweler's rouge were used over the years to scrape away tons of glass and sculpt the disk into a paraboloid within 2/1,000,000 of an inch of perfection. Such patience and care won the day. Operating since 1948, the Hale telescope still remains one of the most useful optical telescopes on Earth, even though its once-record size has since been surpassed. As soon as the glass disk was coated with a few grams of aluminum, just 1,000 atoms thick, to turn it into a reflective mirror, the giant concave surface enabled astronomers to peer farther out into the universe than ever before. It helped reveal that quasars were among the first galactic lights to turn on at the dawn of time.

With the Hale telescope successfully put to work, the Caltech optical lab was eventually converted for another use. Instrumentation was erected to gaze inward, down into the inner workings of the atomic nucleus. But now with the synchrotron removed, part of the space returns to its original purpose: perfecting the technology to once again look outward into the cosmos. This time, though, the business at hand is not devoted to a mirror that will gather light waves. Rather, it is to assist physicists in their quest to place their ears on the fabric of space-time and listen to its distinctive sounds.

At first a few notes will register. In time this should lead to a melody that eventually swells into a lush resounding symphony. And when this happens, astronomers will at last be able to discern the hidden rhythms of the universe.

Coda

This day and age we're living in
Give cause for apprehension
With speed and new invention
And things like fourth dimension
Yet we get a trifle weary
With Mr. Einstein's theory
So we must get down to earth at times
Relax, relieve the tension
And no matter what the progress
Or what may yet be proved
The simple facts of life are such
They cannot be removed
You must remember this
A kiss is just a kiss
A sigh is just a sigh
The fundamental things apply
As time goes by. . . .

—Original introduction to *As Time Goes By*, composed by Herman Hupfeld
for the 1931 Broadway show *Everybody's Welcome* and later used in the movie
Casablanca. (Copyright 1931 by Warner Brothers, Inc. Used by permission.)

Bibliography

Abbott, David (ed.). *Mathematicians.* London: Blond Educational, 1985.

Aglietta, M., et al. "Correlation Between the Maryland and Rome Gravitational-Wave Detectors and the Mont Blanc, Kamioka and IMB Particle Detectors During SN 1987A." *Il Nuovo Cimento,* 106(November 1991):1257-1269.

Allen, B., J. K. Blackburn, A. Lazzarini, T. A. Prince, R. Williams, and H. Yamamoto. "White Paper Outlining the Data Analysis System (DAS) for LIGO I." Document LIGO-M970065-B, June 9, 1997.

Anderson, Christopher. "Divorce Splits LIGO's 'Dysfunctional Family.' " *Science,* 260(May 21, 1993):1063.

Anderson, Christopher. "LIGO Director Out in Shakeup." *Science,* 263(March 11, 1994):1366.

Angel, Roger B. *Relativity: The Theory and Its Philosophy.* Oxford: Pergamon Press, 1980.

Barish, Barry, and Rainer Weiss. "Gravitational Waves Really Are Shifty." *Physics Today,* 53(March 2000):105.

Barish, Barry, and Rainer Weiss. "LIGO and the Detection of Gravitational Waves." *Physics Today,* 52(October 1999):44-50.

Bartusiak, Marcia. "Celestial Zoo." *Omni,* 5(December 1982):106-113.

Bartusiak, Marcia. "Einstein's Unfinished Symphony." *Discover,* 10(August 1989):62-69.

Bartusiak, Marcia F. "Experimental Relativity: Its Day in the Sun." *Science News,* 116(August 25, 1979):140-142.

Bartusiak, Marcia. "Gravity Wave Sky." *Discover,* 14(July 1993):72-77.

Bartusiak, Marcia. "Sensing the Ripples in Space-Time." *Science 85*, 6(April 1985):58-65.

Begelman, Mitchell, and Martin Rees. *Gravity's Fatal Attraction.* New York: Scientific American Library, 1996.

Bell, E. T. *Men of Mathematics.* New York: Simon and Schuster, 1965.

Bender, P., et al. (LISA Study Team). LISA: Laser Interferometer Space Antenna for the Detection and Observation of Gravitational Waves. Pre-Phase A Report, second edition. Garching, Germany: Max-Planck-Institut für Quantenoptik, July 1998.

Blair, David, and Geoff McNamara. *Ripples on a Cosmic Sea.* Reading, Mass.: Helix Books, Addison-Wesley, 1997.

Bromberg, Joan Lisa. *The Laser in America, 1950-1970.* Cambridge, Mass.: MIT Press, 1991.

Burnell, Jocelyn Bell. "The Discovery of Pulsars." In *Serendipitous Discoveries in Radio Astronomy.* Proceedings of Workshop Number 7 held at the National Radio Astronomy Observatory, Green Bank, W. Va., May 4-6, 1983.

Carmeli, Moshe, Stuart I. Fickler, and Louis Witten (eds.). *Relativity: Proceedings of the Relativity Conference in the Midwest.* New York: Plenum Press, 1970.

Chandrasekhar, S. *Eddington.* Cambridge: Cambridge University Press, 1983.

Ciufolini, Ignazio, and John Archibald Wheeler. *Gravitation and Inertia.* Princeton, N.J.: Princeton University Press, 1995.

Cohen, I. Bernard. *Roemer and the First Determination of the Velocity of Light.* New York: Burndy Library, 1944.

Coles, Mark. "How the LIGO Livingston Observatory Got Its Name." *Latest from LIGO Newsletter,* 3(April 1998).

Collins, H. M. *Changing Order: Replication and Induction in Scientific Practice.* Chicago: University of Chicago Press, 1992.

Collins, H. M. "A Strong Confirmation of the Experimenters' Regress." *Studies in History and Philosophy of Science,* 25(1994):493-503.

Committee on Gravitational Physics, National Research Council. *Gravitational Physics: Exploring the Structure of Space and Time.* Washington, D.C.: National Academy Press, 1999.

Coyne, Dennis. "The Laser Interferometer Gravitational-Wave Observatory (LIGO) Project." In *Proceedings of the IEEE 1996 Aerospace Applications Conference,* 4(1996):31-61.

D'Abro, A. *The Evolution of Scientific Thought: From Newton to Einstein.* New York: Dover Publications, 1950.

Damour, T. "Theoretical Aspects of Gravitational Radiation." In *General Relativity and Gravitation: Proceedings of the 14th International Conference.* Singapore: World Scientific, 1997.

Davies, P. C. W. *The Search for Gravity Waves.* Cambridge: Cambridge University Press, 1980.

Davies, P. C. W. *Space and Time in the Modern Universe.* Cambridge: Cambridge University Press, 1977.

Dietrich, Jane. "Realizing LIGO." *Engineering & Science,* 61:2(1998):8-17.

Drever, R. W. P. "Gravitational Wave Astronomy." *Quarterly Journal of the Royal Astronomical Society,* 18(1977):9-27.

Dunnington, G. Waldo. *Carl Friedrich Gauss: Titan of Science*. New York: Exposition Press, 1955.

Dyson, Freeman. "Gravitational Machines." In *Interstellar Communication*. New York: W. A. Benjamin, 1963.

Eddington, A. S. *The Mathematical Theory of Relativity*. Cambridge: Cambridge University Press, 1930.

Eddington, Arthur. *Space, Time and Gravitation*. Cambridge: Cambridge University Press, 1920.

Eddington, Arthur. *The Theory of Relativity and Its Influence on Scientific Thought* (The Romanes Lecture, 1922). Oxford: Clarendon Press, 1922.

Einstein, A. "Die Grundlagen der allgemeinen Relativitätstheorie." *Annalen der Physik*, 49(1916):769-822.

Einstein, Albert. *Relativity: The Special and the General Theory*. New York: Crown Trade Publishers, 1961.

Einstein, A. *Sitzungsberichte der Königlich Preussichen Akademie der Wissenschaften*, (1916):688.

Ellis, George F. R., and Ruth M. Williams. *Flat and Curved Space-Times*. Oxford: Clarendon Press, 1988.

Ezrow, D. H., N. S. Wall, J. Weber, and G. B. Yodh. "Insensitivity to Cosmic Rays of the Gravity Radiation Detector." *Physical Review Letters*, 24(April 27, 1970):945-947.

Faber, Scott. "Gravity's Secret Signals." *New Scientist*, 144(November 26, 1994):40.

Fauvel, John, Raymond Flood, Michael Shortland, and Robin Wilson (eds.). *Let Newton Be!* Oxford: Oxford University Press, 1988.

Fermi, Laura, and Gilberto Bernardini. *Galileo and the Scientific Revolution*. New York: Basic Books, 1961.

Ferrari, V., G. Pizzella, M. Lee, and J. Weber. "Search for Correlations Between the University of Maryland and the University of Rome Gravitational Radiation Antennas." *Physical Review D*, 25(May 15, 1982):2471-2486.

Finn, Lee Samuel. "A Numerical Approach to Binary Black Hole Coalescence." In *General Relativity and Gravitation: Proceedings of the 14th International Conference*. Singapore: World Scientific, 1997.

Fisher, Arthur. "The Tantalizing Quest for Gravity Waves." *Popular Science*, 218(April 1981):88-94.

Flam, Faye. "A Prize for Patient Listening." *Science*, 262(October 22, 1993):507.

Flam, Faye. "Scientists Chase Gravity's Rainbow." *Science*, 260(April 23, 1993):493.

Florence, Ronald. *The Perfect Machine: Building the Palomar Telescope*. New York: HarperCollins, 1994.

Folkner, William M. (ed.). *Laser Interferometer Space Antenna: Second International LISA Symposium on the Detection and Observation of Gravitational Waves in Space*. AIP Conference Proceedings 456. Woodbury, N.Y.: American Institute of Physics, 1998.

Fölsing, Albrecht. *Albert Einstein: A Biography*. New York: Viking, 1997.

Fomalont, E. B., and R. A. Sramek. "Measurements of the Solar Gravitational Deflection of Radio Waves in Agreement with General Relativity." *Physical Review Letters*, 36(June 21, 1976):1475-1478.

Forward, Robert L. "Wideband Laser-Interferometer Gravitational-Radiation Experiment." *Physical Review D*, 17(January 15, 1978):379-390.

Franklin, Allan. "How to Avoid the Experimenters' Regress." *Studies in History and Philosophy of Science*, 25(1994):463-491.

Friedman, Michael. *Foundations of Space-Time Theories: Relativistic Physics and Philosophy of Science.* Princeton, N.J.: Princeton University Press, 1983.

Garwin, Richard L. "Detection of Gravity Waves Challenged." *Physics Today*, 27(December 1974):9-11

Garwin, R. L. "More on Gravity Waves." *Physics Today*, 28(November 1975):13.

Garwin, Richard L., and James L. Levine. "Single Gravity-Wave Detector Results Contrasted with Previous Coincidence Detections." *Physical Review Letters*, 31(July 16, 1973):176-180.

Gertsenshtein, M. E., and V. I. Pustovoit. "On the Detection of Low Frequency Gravitational Waves." *Soviet Physics JETP*, 16(February 1963):433-435.

Giazotto, Adalberto. "Interferometric Detection of Gravitational Waves." *Physics Reports*, 182(November 1989):365-424.

Gibbons, G. W., and S. W. Hawking. "Theory of the Detection of Short Bursts of Gravitational Radiation." *Physical Review D*, 4(October 15, 1971):2191-2197.

Gillispie, Charles Coulston (ed.). *Dictionary of Scientific Biography.* New York: Charles Scribner's Sons, 1972.

Glanz, James. "Gamma Blast from Way, Way Back." *Science*, 280(April 24, 1998):514.

Gorman, Peter. *Pythagoras: A Life.* London: Routledge & Kegan Paul, 1979.

"Gravitating Toward Einstein." *Time*, 93(June 20, 1969):75.

"Gravitational Waves Detected." *Science News*, 95(June 21, 1969):593-594.

"Gravitational Waves Detected." *Sky & Telescope*, 38(August, 1969):71.

"Gravity Waves Slow Binary Pulsar." *Physics Today*, 32(May, 1979):19-20.

Hall, Tord. *Carl Friedrich Gauss.* Cambridge, Mass.: MIT Press, 1970.

Hariharan, P. *Optical Interferometry.* Sydney: Academic Press, 1985.

Hawking, S. W., and W. Israel (eds.). *General Relativity: An Einstein Centenary Survey.* Cambridge: Cambridge University Press, 1979.

Hilts, Philip J. "Last Rites for a 'Plywood Palace' That Was a Rock of Science." *The New York Times* (March 31, 1998):C4.

Hinckfuss, Ian. *The Existence of Space and Time.* Oxford: Clarendon Press, 1975.

Hoffmann, Banesh. *Albert Einstein: Creator and Rebel.* New York: The Viking Press, 1972.

Hoffmann, Banesh. *Relativity and Its Roots.* New York: W. H. Freeman, 1983.

Hulse, Russell A. "The Discovery of the Binary Pulsar." *Reviews of Modern Physics*, 66(July 1994):699-710.

Jammer, Max. *Concepts of Space.* Cambridge, Mass.: Harvard University Press, 1954.

Kepler, Johannes. *The Harmony of the World.* Philadelphia: American Philosophical Society, 1997.

Kimball, Robert (ed.). *The Complete Lyrics of Cole Porter.* New York: Alfred A. Knopf, 1983.

Kleppner, Daniel. "The Gem of General Relativity." *Physics Today*, 46(April 1993):9-11.

Laser Pioneer Interviews. Torrance, Calif.: High Tech Publications, 1985.

Lee, M., D. Gretz, S. Steppel, and J. Weber. "Gravitational-Radiation-Detector Observations in 1973 and 1974." *Physical Review D*, 14(August 15, 1976):893-906.

Levine, James L., and Richard L. Garwin. "Absence of Gravity-Wave Signals in a Bar at 1695 Hz." *Physical Review Letters*, 31(July 16, 1973):173-176.

Lubkin, Gloria B. "Experimental Relativity Hits the Big Time." *Physics Today*, 23(August 1970):41-44.

Lubkin, Gloria B. "Weber Reports 1660-Hz Gravitational Waves from Outer Space." *Physics Today*, 22(August 1969):61-62.

Machamer, Peter K., and Robert G. Turnbull (eds.). *Motion and Time/Space and Matter: Interrelations in the History of Philosophy and Science*. Columbus: Ohio State University Press, 1976.

Marshall, Eliot. "Garwin and Weber's Waves." *Science*, 212(May 15, 1981):765.

Mather, John C., and John Boslough. *The Very First Light*. New York: Basic Books, 1996.

Michelson, Peter F., John C. Price, and Robert C. Taber. "Resonant-Mass Detectors of Gravitational Radiation." *Science*, 237(July 10, 1987):150-157.

Minkowski, H. "Das Relativitätsprinzip." *Jahresbericht der Deutschen Mathematiker-Vereinigung*, 24:372-382.

Misner, C.W., K. S. Thorne, and J. A. Wheeler. *Gravitation*. San Francisco: W. H. Freeman, 1973.

Monastyrsky, Michael. *Riemann, Topology, and Physics*. Boston: Birkhäuser, 1987.

Moss, G. E., L. R. Miller, and R. L. Forward. "Photon-Noise-Limited Laser Transducer for Gravitational Antenna." *Applied Optics*, 10(November 1971):2495-2498.

Naeye, Robert. "To Catch a Gravity Wave." *Discover*, 14(July 1993):76.

Norton, John. "How Einstein Found His Field Equations, 1912-1915." In *Einstein and the History of General Relativity*, Don Howard and John Stachel (eds.). Boston: Birkhäuser, 1989.

Overbye, Dennis. "Testing Gravity in a Binary Pulsar." *Sky & Telescope*, 57(February 1979):146.

Pais, Abraham. '*Subtle Is the Lord . . .*': *The Science and the Life of Albert Einstein*. Oxford: Oxford University Press, 1982.

Park, David. *The How and the Why*. Princeton, N.J.: Princeton University Press, 1988.

Perkowitz, Sidney. *Empire of Light*. New York: Henry Holt, 1996.

Perryman, Michael. "Hipparcos: The Stars in Three Dimensions." *Sky & Telescope*, (June 1999):40-50.

Pirani, F. A. E. "On the Physical Significance of the Riemann Tensor." *Acta Physica Polonica*, 15(1956):389-405.

Pound, R. V., and G. A. Rebka, Jr. "Apparent Weight of Photons." *Physical Review Letters*, 4(April 1, 1960):337-340.

Preparata, Giuliano. "Joe Weber's Physics." In *QED Coherence in Matter*. Singapore: World Scientific, 1995.

Press, William H., and Kip S. Thorne. "Gravitational-Wave Astronomy." *Annual Review of Astronomy and Astrophysics* (1972):74.

Preston, Richard. *First Light: The Search for the Edge of the Universe*. New York: Atlantic Monthly Press, 1987.

Raine, Derek J., and Michael Heller. *The Science of Space-Time*. Tucson: Pachart Publishing House, 1981.

Salmon, Wesley C. *Space, Time, and Motion*. Minneapolis: University of Minnesota Press, 1980.

Sanders, J. H. *The Velocity of Light.* Oxford: Pergamon Press, 1965.

Saulson, Peter R. *Fundamentals of Interferometric Gravitational Wave Detectors.* Singapore: World Scientific, 1994.

Saulson, Peter R. "If Light Waves Are Stretched by Gravitational Waves, How Can We Use Light as a Ruler to Detect Gravitational Waves?" *American Journal of Physics*, 65(June 1997):501-505.

Schilpp, Paul A. (ed.). *Albert Einstein: Philosopher-Scientist.* New York: Harper Torchbooks, 1959.

Schwinger, Julian. *Einstein's Legacy.* New York: Scientific American Books, 1986.

Shapiro, Irwin I. "A Century of Relativity." In *More Things in Heaven and Earth: A Celebration of Physics at the Millennium,* Benjamin Bederson, ed. New York: Springer-Verlag, 1999.

Shapiro, Irwin I. "Fourth Test of General Relativity." *Physical Review Letters*, 13(December 28, 1964):789-791.

Shapiro, Irwin I., et al. "Fourth Test of General Relativity: New Radar Result." *Physical Review Letters*, 26(May 3, 1971):1132-1135.

Shapiro, Irwin I., et al. "Fourth Test of General Relativity: Preliminary Results." *Physical Review Letters*, 20(May 27, 1968):1265-1269.

Shapiro, Stuart L., Richard F. Stark, and Saul A. Teukolsky. "The Search for Gravitational Waves." *American Scientist*, 73(May/June 1985):248-257.

Shaviv, G., and J. Rosen (eds.). *General Relativity and Gravitation: Proceedings of the Seventh International Conference (GR7).* New York: John Wiley & Sons, 1975.

Silk, Joseph. "Black Holes and Time Warps: Einstein's Outrageous Legacy." *Science*, 264 (May 13, 1994):999.

Sklar, Lawrence. *Space, Time, and Spacetime.* Berkeley: University of California Press, 1974.

Smart, J. J. C. (ed.). *Problems of Space and Time.* New York: Macmillan, 1964.

Sobel, Michael I. *Light.* Chicago: University of Chicago Press, 1987.

Taubes, Gary. "The Gravity Probe." *Discover*, 18 (March 1997):62.

Taylor, Edwin F., and John Archibald Wheeler. *Spacetime Physics.* New York: W. H. Freeman, 1966.

Taylor, Joseph H., Jr. "Binary Pulsars and Relativistic Gravity." *Reviews of Modern Physics*, 66(July 1994):711-719.

Thomsen, Dietrick. "Gaining on Gravity Waves." *Science News*, 98(July 11, 1970):44-45.

Thomsen, Dietrick. "Trying to Rock with Gravity's Vibes." *Science News*, 126(August, 4, 1984):76-78.

Thorne, Kip. *Black Holes and Time Warps: Einstein's Outrageous Legacy.* New York: W. W. Norton, 1994.

Thorne, Kip S. "Gravitational-Wave Research: Current Status and Future Prospects." *Reviews of Modern Physics*, 52(April 1980):285-297.

Torretti, Roberto. *Relativity and Geometry.* Oxford: Pergamon Press, 1983.

Tourrene, Philippe. *Relativity and Gravitation.* Cambridge: Cambridge University Press, 1997.

Travis, John. "LIGO: A $250 Million Gamble." *Science*, 260(April 30, 1993):612-614.

Trimble, Virginia, and Joseph Weber. "Gravitational Radiation Detection Experiments with Disc-Shaped and Cylindrical Antennae and the Lunar Surface

Gravimeter." *Annals of the New York Academy of Sciences*, 224(December 14, 1973):93-107.

Tyson, J. A. "Gravitational Radiation." *Annals of the New York Academy of Sciences*, 224(December 14, 1973):74-92.

Tyson, J. Anthony. Testimony before the Subcommittee on Science of the Committee on Science, Space, and Technology, United States House of Representatives, March 13, 1991.

Vessot, R. F. C., and M. W. Levine. "A Test of the Equivalence Principle Using a Space-Borne Clock." *General Relativity and Gravitation*, 10(1979):181-201.

Vessot, R. F. C., et al. "Test of Relativistic Gravitation with a Space-borne Hydrogen Maser." *Physical Review Letters*, 45(December 29, 1980):2081-2084.

Wald, Robert M. *Space, Time, and Gravity.* Chicago: University of Chicago Press, 1992.

Waldrop, M. Mitchell. "Of Politics, Pulsars, Death Spirals—and LIGO." *Science*, 249(September 7, 1990):1106.

Weber, J. "Anisotropy and Polarization in the Gravitational-Radiation Experiments." *Physical Review Letters*, 25(July 20, 1970):180-184.

Weber, J. "Detection and Generation of Gravitational Waves." *Physical Review*, 117(January 1, 1960):306-313.

Weber, Joseph. "The Detection of Gravitational Waves." *Scientific American* 224 (May 1971):22-29.

Weber, J. "Evidence for Discovery of Gravitational Radiation." *Physical Review Letters*, 22(June 16, 1969):1320-1324.

Weber, J. "Gravitational Radiation Detector Observations in 1973 and 1974." *Nature*, 266(March 17, 1977):243.

Weber, Joseph. "Gravitational Waves." *Physics Today*, 21(April 1968).34-39.

Weber, J. "Gravitons, Neutrinos, and Antineutrinos." *Foundations of Physics*, 14(December 1984):1185-1209.

Weber, Joseph. "How I Discovered Gravitational Waves." *Popular Science*, 200(May 1972):106+.

Weber, Joseph. "Weber Replies." *Physics Today*, 27(December 1974):11-13.

Weber, J. "Weber Responds." *Physics Today*, 28(November 1975):13+.

Weber, J., and T. M. Karade (eds.). *Gravitational Radiation and Relativity.* Singapore: World Scientific, 1986.

Weber, J., and B. Radak. "Search for Correlations of Gamma-Ray Bursts with Gravitational-Radiation Antenna Pulses." *Il Nuovo Cimento*, 111B(June 1996):687-692.

Weiss, Rainer. "Gravitational Radiation." In *More Things in Heaven and Earth: A Celebration of Physics at the Millennium*, Benjamin Bederson, ed. New York: Springer-Verlag, 1999.

Weiss, Rainer. "Gravitation Research." Quarterly Progress Report Number 105. Research Laboratory of Electronics, Massachusetts Institute of Technology, (1972):54-76.

Westfall, Richard S. *The Life of Isaac Newton.* Cambridge: Cambridge University Press, 1993.

Wheeler, John Archibald, and Kenneth Ford. *Geons, Black Holes, and Quantum Foam.* New York: W. W. Norton, 1998.

Will, Clifford M. "The Confrontation Between Gravitation Theory and Experiment." In *General Relativity: An Einstein Centenary Survey*. Cambridge: Cambridge University Press, 1979.

Will, Clifford M. "General Relativity at 75: How Right Was Einstein?" *Science*, 250(November 9, 1990):770-776.

Will, Clifford M. "Gravitational Radiation and the Validity of General Relativity." *Physics Today*, 52(October 1999):38-43.

Will, Clifford M. *Was Einstein Right?* New York: Basic Books, 1986.

Williams, L. Pearce (ed.). *Relativity Theory: Its Origins and Impact on Modern Thought*. Huntington, N.Y.: Robert E. Krieger, 1979.

Index

A

Abell 2218, 51–52
Acceleration, and gravity, 39–41, 52–53
Adelberger, Eric, 64
AIGO (Australian International
    Gravitational Observatory),
    187–88
Albert Einstein Institute, 185
Allegro, 106–7, 109
Altair, 71
American Institute of Physics, 103
American Physical Society, 37
Anderson, Philip, 141
Andromeda galaxy, 109
Apollo missions, 92, 217
Aquila constellation, 79
Archytas, 12
Arecibo Observatory, 76–78, 81, 83
Argonne National Laboratory, 94, 104
Aristotle, 12, 13, 176

Armstrong, John, 216
As Time Goes By, 231
Atomic clocks, 54–55, 64, 118, 120–21,
    215
Atomic fountain, 121
Auriga, 108
Australian National University, 187

B

Baade, Walter, 72
Bar detectors, 89
    Allegro, 106–7, 109
    amplifiers, 100, 108
    Auriga, 108
    bar materials, 97, 107, 108–9
    calibration, 101
    Earth as resonant bar, 92, 97–98
    Explorer, 108, 109
    frequency range, 97, 107–8, 125
    groups developing, 96–97, 103–5

interference/error sources, 97,
101–4, 106–7, 109
interpretation of response, 9–10,
103, 113
isolation and suspension system, 93
limitations, 125
linked piezoelectric transducers, 104
long-distance masses, 106
Nautilus, 108, 109
network, 108–9
Niobe, 108
piezoelectric crystal receivers, 92, 93
principle, 9–10, 93
second-generation, 109
sensitivity, 105, 107, 109, 114
size of, 93, 100, 125
spherical, 108–9, 176–77
superconducting instrumentation,
107
supercooled bars, 98, 106–7, 108,
125, 129, 180
type of signal registered by, 98–99
Bardon, Marcel, 139
Barish, Barry, 68, 146–47, 161, 167,
190
Basov, Nikolai, 92
BATSE (Burst and Transient Source
Experiment), 114
Dickedoiff, David, 61
Bell Burnell, Jocelyn, 70–71, 72, 86
Bell Laboratories, 96, 100
Bender, Peter, 216–19
BeppoSAX satellite, 205
Berkeley, George, 16
Bertotti, Bruno, 210
Bhawal, Biplab, 168
Big Bang, 3, 63, 132, 208, 226
Billing, Heinz, 180
Billingsley, GariLynn, 155–56
Binary star systems
gravitational waves from, 81–83, 93,
177, 181, 192
pulsars, 75, 78–86, 181
white dwarf collisions, 225–26
x-ray, 192, 197

Black Death, 13
Black holes, 3, 6, 118, 195
colliding, 88, 109, 153, 171, 192–93,
198, 198–203, 222
Cygnus X-1, 197
evidence of, 192, 196, 198, 224
formation/source, 200, 204, 205
gamma-ray bursts, 114
general theory of relativity and,
59–62, 201–2
gravitational field, 53
gravitational waves, 68, 88, 93, 96,
169, 171, 192–93, 197, 199,
201–2, 214
mass, 57, 171, 200
neutron star merger, 171
NSF Grand Challenge, 201–2
origin of term, 62
properties, 199–200, 225
rotation, 56, 199–200, 225
signal characteristics, 153, 200
supermassive, 57, 201, 214, 223–25
Bohr, Niels, 57, 59, 112
Bolyai, János, 20
Bose speaker, 118
Braginsky, Vladimir, 64, 97, 100, 131,
133, 218
Brans, Carl, 65
Brans-Dicke theory, 63–66, 97
Brillet, Alain, 178
Bronx High School of Science, 75
Brookhaven National Laboratory, 168
Brownian motion, 31
Browning, Robert, 1

C

Calculus, invention of, 13, 16
California Institute of Technology
(Caltech), 52, 63, 95, 184, 194
Bridge Building, 157
general relativity research, 131, 195
gravity wave program, 132, 134–36
138–39, 144, 155, 228

laser interferometer prototype,
  133–36, 138, 142, 155, 161, 165
LIGO, 2, 138, 143, 147, 155, 157,
  165, 167
millimeter-wave radio astronomy
  array, 140, 146
optical lab, 229–30
Palomar telescope, 141, 144
pulsar research, 72
quasar research, 195
Synchrotron Laboratory, 228
Cambridge University (England), 70
Cardiff University, 102
Carleton College, 83
Case School of Applied Science, 27
Cassini mission, 216
Celestial coordinates, 79
Celestial distance measurements, 204,
  217
Center for Relativity, 202
CERN (European Center for Nuclear
  Research), 108, 110, 178
Cerro Tololo Observatory, 100
Chandra X-ray Observatory, 196, 219
Chandrasekhar, Subrahmanyan, 196
Chapman, Philip, 125, 128
Chomsky, Noam, 117
COBE (Cosmic Background Explorer)
  satellite, 132, 138, 173, 208–9
Cold fusion, 96
Coles, Mark, 3
Collins, Harry, 102
Color, theory of, 13
Colwell, Rita, 3
Compton Gamma-Ray Observatory, 114
Computer applications
  black-hole collision simulation,
    202
  gravitational wave analysis, 93,
    165–66, 202
  LIGO simulations, 163–64
  orbital simulation, 76
  pulsar detection, 77–79, 85
Conservation of angular momentum,
  57, 73

Contact, 194
Cooper Union, 76
Copernicus, 37–38
Cornell University, 81
Corning, 156
Cosmic microwave background, 63,
  132, 153, 170, 208, 226
Cosmic rays, 109, 129, 192
Cosmic strings, 195, 209, 227
Crab Nebula, 79
CSIRO (Commonwealth Scientific and
  Industrial Research
  Organization), 156
Cygnus X-1, 197

D

Danzmann, Karsten, 183, 185, 186, 219
Data analysis, 164–69, 185
Deep Space Network, 214
Democritus, 12
Dicke, Robert, 63–66, 67, 121, 122, 126,
  167, 195, 196, 217
Disturbed storm-time factor, 101
Domain walls, 209
Douglass, David, 100, 102–3, 104
Dragging, 55–57, 82, 204, 222
Drever, Ronald
  Caltech gravity wave program, 133,
    134–36, 138, 144–45, 155,
    228–29
  expertise, 133
  Glasgow gravity wave program, 95,
    104–5, 129–30, 133, 134, 181
  gravitational redshift detectors, 129
  inertia experiments, 129
  laser power recycling, 135, 182
  LIGO construction, 133, 138, 140,
    141, 144–45, 147, 160, 169
  personal characteristics, 140, 145
  space-based detectors, 218
  on Weber's findings, 106
  Weiss and, 140
Dyson, Freeman, 90, 195, 203

E

Earth
  gravitational radiation, 68
  internal gravity gradients, 229
  magnetic field, 73, 101
  as resonant bar, 92, 98
  tides, 154
Eddington, Arthur, 37–38, 48–49, 50,
    51, 69
Edgerton, Harold, 118
Einstein, Albert. *See also* General theory
    of relativity; Special theory of
    relativity
  appearance and demeanor, 29
  collaborators, 41–42, 43, 47, 57
  as college student, 28, 30, 34
  on common sense, 38
  conversations with Newton, 16, 46
  and electromagnetism, 28–29, 30
  gravitational radiation concept,
    68–69
  professional life, 30, 39, 59, 90
  speed of light, 23, 28
  superstardom/mystique, 4, 49–50,
    141, 143, 231
  thought experiments, 23, 39–41
  youth, 23, 30, 39
Electricity, 25–26
Electromagnetic radiation, 26, 73, 192,
    193
Electromagnetism, 8
  Faraday's experiments, 26
  Maxwell's field equations, 26, 29,
    89
  Newtonian mechanics and, 30
  Ørsted's experiments, 25–26
Electron Tube Research Conference,
    91
Elements, The, 17–18
Elieson, Steve, 157
Ether, luminiferous
  light propagation in, 25, 32–34
  motion of Earth through, 27
  wind, 27, 28, 34

Ettore Majorana Center for Scientific
    Culture, 131
Euclid, 17–18
European Space Agency (ESA), 50,
    214, 219, 220
Event horizon, 60, 210
Explorer (bar detector), 108, 109

F

Fabry-Perot interferometer, 130,
    141
Fairbank, William, 95, 98, 107
Faller, James, 217, 218–19
Faraday, Michael, 26
Fifth Cambridge Conference on
    Relativity, 103
Finn, Sam, 192, 203
FitzGerald, George, 28, 32
Five College Radio Astronomy
    Observatory, 74
Florence, Ronald, 144
Folkner, William, 219
Forward, Robert L., 92–93, 125,
    126–28, 165, 181
Fowler, William, 195
Franklin, Cecil, 159
Frequency range
  bar detectors, 97, 108, 125
  laser interferometer detectors, 126,
    153, 182
  LIGO, 153, 222
  space-based detectors, 215, 216,
    222
Fulton, Robert, 2
Fused silica mirrors, 134, 156, 170,
    181, 188

G

Galaxies
  clusters, 51–52
  colliding, 223, 225
  gravitational lensing, 51–52

FSC 10214+4724, 52
MCG 6–30–15, 210
Galilei, Galileo, 8, 13, 15, 24, 92,
        175–76
Galileo probe, 214
Gamma rays
    bursts, 114–15, 166, 205
    detectors, 114, 189
    frequency shift, 54
Gamow, George, 194
Garwin, Richard, 102, 139
Gauss, Carl Friedrich, 18–19, 20, 21
General Optics, 156
General theory of relativity
    alternative theories, 63–66
    applications, 6
    and black holes, 60–62, 202
    corrections to calculations, 84
    and dragging (gravitomagnetism),
        55–57
    equation, 43–44, 69
    evidence for, 5, 6
    gravity and, 39, 41–42, 43–44, 46,
        50–53, 69, 202
    joke, 53
    and light deflection, 41, 44, 47–49,
        50–52, 54
    LURE test, 217
    main significance of, 47
    Mercury's orbit and, 42–43, 44, 47,
        76, 82
    Poincaré and, 28
    problem in interpreting results,
        69–70
    pulsar experiments, 81–86
    radar experiments, 51, 62
    rebirth of, 62
    reference frame, 42, 45–46
    Riemannian curvature and, 21–22,
        41, 45
    and space-time concept, 4, 44–45
    starlight deflection experiments,
        48
    Strong Equivalence Principle, 217
    technology for testing, 6, 121, 213
    time in gravitational field, 52–53

German University in Prague, 39
GEO, 183–84
GEO 600, 176, 184–87, 219
Geometry
    Euclidean, 17–18
    imaginary, 20
    negative curvature, 20, 21
    non-Euclidean, 18–22, 41
    positive curvature, 20–21
Gertsenshtein, Mikhail E., 124
Giazotto, Adalberto, 177, 178–80
Gibbons, Gary, 98
Global Positioning System, 6, 55
Globular clusters, 200
Goldenberg, H. Mark, 66
Grail Project, 109–10
Gran Sasso, 146
Gravimeter, 92, 121
Gravitation, 6, 82, 132, 142, 197
Gravitational constant, 122
Gravitational fields
    black holes, 53
    clocks in, 41, 52–53
Gravitational lenses, 6, 51–52, 113, 210
Gravitational mass, equivalence to
        inertial mass, 64
Gravitational Physics, 201
Gravitational redshift, 54–55, 58, 121,
        129
Gravitational wave detectors. See also
        Bar detectors; Laser
        interferometers; LIGO
    first-detection ground rules,
        188–90
    international network of, 176–77,
        184–90
Gravitational waves
    alternative theories, 63–66, 122
    audio frequency range, 165, 192,
        199–200, 204, 225
    background (primordial), 208–11
    from binary star systems, 82, 96,
        177, 181, 192, 203–5
    from black holes, 68, 88, 93, 96, 109,
        169, 171, 192, 197, 199, 225
    corrections to calculations, 84

defined, 3, 67–68
detection, 68, 81–84. *See also*
    Gravitational wave detectors
Einstein's concept, 68–69
electromagnetic radiation
    compared, 192
evidence of, 8, 70–71, 100, 110–11
information conveyed by, 3, 204,
    205–6, 208–9, 225
and light waves, 123n, 192
from neutron stars, 82, 93, 95, 96,
    98, 108, 166, 171, 192, 197,
    203–5
peak-to-peak span, 165, 179,
    214
properties, 7, 9–10, 69, 90, 122, 192,
    193
from pulsars, 81–84, 95–96, 109,
    166, 168, 177
from Sanduleak–69° 202, 6–7
strength of (strain), 68, 69, 87–88,
    98–99, 100, 131, 137, 161, 169,
    184, 188
from supernovas, 68, 69, 88, 90, 95,
    96, 105, 108, 110, 153, 166, 171,
    177, 178, 192–93, 198, 205–7
symmetrical, 206
tracing signal source, 95–96, 103,
    126, 177, 187
Gravitomagnetism, 55–57
Graviton, 178
Gravity
    and acceleration, 39–41, 53
    force of, 13–14, 52–53
    Galileo's experiments, 175–76
    general theory of relativity and, 39,
        41–42, 43–44, 46, 50–53, 69,
        202
    and light, 39–41, 44, 45, 47–49,
        50–52, 54
    Mach's principle, 64–65, 128
    Newton's laws of, 5, 13–14, 15–16,
        17, 41–42, 45, 48
    and space-time curvature, 41,
        43–46

Gravity Probe A, 56
Gravity Probe B, 56–57
Gravity waves. *See* Gravitational
    waves
Gretarsson, Andri, 172
Grossman, Marcel, 41–42
Gyroscope experiment, 56–57

H

Hale, George, 229
Hale telescope, 229–30
Halley, Edmond, 13
Hamilton, William, 107
Hanford Nuclear Reservation, 149–50,
    153
Harpedonaptai, 17
Harvard-Smithsonian Center for
    Astrophysics, 54
Harvard University, 70, 71, 129
    Jefferson Physical Laboratory, 54
Harwit, Martin, 170
Haverford College, 74
Hawking, Stephen, 98, 210
Haystack observatory, 51
Heraeus, 156
Hertz, Heinrich, 27, 89
Hewish, Antony, 70, 71, 86
Hinckfuss, Ian, 11
Hipparcos satellite, 50
Hoffmann, Banesh, 47
Hough, James, 130, 184, 185, 219
Hoyle, Fred, 195
Hubble, Edwin, 58
Hubble Space Telescope, 159, 173,
    219
Hughes Aircraft Research Laboratories,
    92–93
Hughes-Drever experiments,
    128–29
Hughes Laser Interferometer
    Gravitational Radiation
    Antenna, 125–28
Hughes, Vernon, 129

Hulse, Russell, 75–82
Hulse-Taylor binary, 78–80, 81–82,
    83–84, 85, 168, 192, 203
Hydrogen bomb, 58, 102
Hydrogen maser clock, 54–55

I

IBM, 96, 102
Inertia, 37–38, 64, 129
Inflationary universes, 6
Institute for Advanced Study, 57, 90
Instituto Nazionale di Fisica
    Nucleare (INFN), 110, 177, 1
    78, 179
Io, light speed experiments, 24, 25
Isaacson, Richard, 133, 139
Isolation and suspension systems
    fused silica wires, 186
    for laser interferometers, 130,
        134–35, 179–80, 181, 186
    for LIGO, 157–58, 170, 172
    magnetic levitation, 228
    for resonant bar antennas, 93
    VIRGO super attenuator, 179

J

Jansky, Karl, 96
Jet Propulsion Laboratory (JPL), 140,
    216, 219, 220
Johnson, Warren, 107
Joint Institute for Laboratory
    Astrophysics, 97, 130, 217

K

Kafka, Peter, 104
Keck telescopes, 52, 173
Kepler, Johannes, 191–92, 206
Kinematics, 13
Krauss, Lawrence, 208

L

Landau, Lev, 72
Large Magellanic Cloud, 6–7, 110, 136,
    206
Laser Gravitational-wave Observatory
    in Space (LAGOS), 219
Laser gyroscopes, 135
Laser Interferometer Space Antenna.
    See LISA
Laser interferometers, 98. See also LIGO;
    Space-based interferometers
    AIGO, 187–88
    alignment, 127
    Caltech prototype, 133–36, 138, 142,
        155, 161, 165
    frequency range, 126, 182
    GEO 600, 176, 184–87
    German prototype, 181–84
    Hughes Laser Interferometer
        Gravitational Radiation
        Antenna, 123–28
    laser power, 135, 181
    mirrors/test mass, 134–36, 181, 187,
        188
    MIT prototype, 136, 138–39, 181
    noise/interference sources, 126, 127,
        130, 141, 181, 182, 187
    NSF funding, 134
    observatory design, 125–26, 134
    power recycling, 135, 161, 182
    principle, 123, 186
    prototype, 126–29
    sensitivity, 124, 127–28, 132, 135,
        136, 181, 182–83, 186
    setup, 123–25
    signal recycling, 186, 188
    size of instruments, 133, 176, 177,
        181, 182–83, 185, 187
    stabilization, 130, 134, 136, 179, 181,
        186
    strain, 135, 137, 184, 188
    TAMA 300, 187
    timing of events, 127

vacuum systems, 134, 185–86
VIRGO, 176–80, 186, 222
Laser Interferometer Gravitational-
    Wave Observatory. *See* LIGO
Lasers, 64, 91, 97–98
    amplification for space, 220–21
    argon ion, 160–61
    power recycling, 135–36, 161, 182
    solid-state infrared, 161, 162
Lazzarini, Albert, 164, 165, 166
Leaning tower of Pisa, 175–76
Legnaro National Laboratories, 108
Leibniz, Gottfried Wilhelm, 16
Lense, Josef, 55
Lense-Thirring effect, 55–57
Levine, James, 102
Levine, Judah, 97
Libbrecht, Kenneth, 65, 66, 167, 169–70
Light. *See also* Speed of light; Stars
    mass of, 34
    particles (photons), 30, 161
    propagation in luminiferous ether,
      25
    visible, 26–27
Light waves
    gravity and, 5, 39–41, 44, 45, 47–49,
      51–52, 54
    gravitational waves and, 123n, 192
    redshift, 54, 58
LIGO, 115
    Barish and, 146–47, 160–61, 167
    beam tubes, 158–60, 227
    blue-ribbon panels, 139, 142
    Caltech-MIT collaboration, 138
    comparison of findings, 153–54
    control system, 161–63
    costs/budget, 141, 146, 154–55, 157,
      165
    criticisms/opposition, 140, 141–42,
      173
    data analysis, 164–69
    diagnostic probes, 164
    distance between observatories,
      152–53
    Drever and, 133, 138, 140, 141,
      144–45, 160, 169, 228

event tracking, 164–65
expected rates of detections, 171
feasibility study, 138
frequencies, 153, 222
initiation and shakedown, 173, 227
instrumentation, 2, 9
interferometers, 141, 143, 151–52,
    153, 161–63
internal conflicts, 140, 144–45, 147
lasers, 152, 160–61, 162, 170, 193
length of arms, 152, 167, 177, 185,
    186, 203
Louisiana (Livingston) facility, 1–2,
    8, 117, 142–43, 151, 152, 158–59,
    173
mirrors, 154, 155–57, 163, 165, 170,
    172
noise/interference, 152–53, 154, 156,
    157–58, 159–60, 161, 163, 164,
    169, 170, 172
NSF/federal funding, 137–40, 141,
    142–43, 146, 155, 189–90
optics laboratory, 157
principle, 9–10
public education, 151
Scientific Collaboration, 155, 169
second-generation instrumentation,
    169–73, 203, 206
sensitivity, 142, 152, 153, 161, 169,
    186
simulations, 163–64
site characteristics, 152–53
slogan, 3
staffing, 143
strain, 161, 169, 188
suspension and seismic isolation,
    157–58, 170, 172
tone, 157, 165
Thorne and, 138, 139–40, 169,
    193–94
types of events detectable, 153, 166,
    168–69, 171
vacuum system, 117, 151, 158–60
Vogt and, 140–41, 143, 145–46,
    147
volume of observable space, 203

Washington state (Hanford) facility, 8, 117, 146, 149–51, 152–53, 173, 227
Weiss and, 117, 119, 137–40, 141, 159, 173
LINE, 219
Linsay, Paul, 138
LISA, 219–24, 225–26
Little Green Men (LGM), 71
Livingston, Edward, 2
Lobachevsky, Nikolai, 19–20, 21
Lock-in amplifier, 64, 65
Lorentz, Hendrik, 28, 31, 32, 58
Louisiana State University, 97, 107, 108, 129
Lucretius, 12
Lunar Ranging Experiment (LURE), 217, 218

M

Mach, Ernst, 55, 64, 128
Mach's principle, 64–65, 128
Magie, William, 37
Magnetic coils, 154
Magnetic fields
    Earth's, 73, 101
    and gravitational wave detection, 101
    strength of, 73
    from thunderstorms, 154
Magnetic monopoles, 146, 190, 209
Magnetism, 25–26
Majid, Walid, 167–68
Manhattan Project, 58, 59
Mars, 51
Mars Observer, 214
Masers, 64, 91
Mass
    of black holes, 57, 171, 200, 201
    gravitational, 64
    of light, 34
    speed of light and, 34
Mass-energy, 44
Massachusetts Institute of Technology

(MIT), 2, 4, 125, 131, 171, 193
    Building 20, 117–18, 120, 121
    laser interferometry prototype, 136, 138–39, 181
    LIGO, 139, 143, 161
    Research Laboratory for Electronics, 136
Matter, 11, 193
Matzner, Richard, 202
Max-Planck-Institut für Physik und Astrophysik, 180
Max-Planck-Institut für Quantenoptik, 182, 183, 184
Maxwell, James Clerk, 26, 27
Maxwell's field equations, 26, 29, 89
McDonald Observatory, 217
Meers, Brian, 186
Mercury
    orbital motion, 42–43, 44, 47, 76, 82
    perihelion, 42, 44, 65
    radar signal delays, 51
Metrical field, 22, 45
Michelson, Albert A., 27–28, 34, 151
Michelson interferometer, 27–28, 123, 130, 141, 151–52
Microwave energy, 91
Microwave radiometer, 64
Milky Way, 8, 74, 78, 96, 105, 153, 187, 200, 224
Miller, Larry, 126
Minkowski, Hermann, 34–35, 41
Minuteman missile program, 195
Mirrors
    fused silica, 134, 156, 170, 181, 188
    LIGO design, 154, 155–58, 163, 165, 170, 172
    sapphire, 170, 188
    suspension systems, 181
    supercooled, 187
    supermirrors, 182
Misner, Charles, 6, 82, 197
Monopoles, 146–47
Montagu, Ashley, 53
Montana State University, 63
Moon (Luna), gravimeter on, 92

Morley, Edward, 27–28, 34, 151
Morrison, Philip, 103
Moscow State University, 97
Moss, Gaylord, 126–27
Mount Wilson telescope, 229
Muon particles, 34
Music of the spheres, 191–92, 193

N

NASA Goddard Institute for Space
     Studies, 61–62
National Aeronautics and Space
     Administration (NASA), 56, 113,
     114–15, 125, 128, 132, 219
     Binary Neutron Star Grand
         Challenge, 204
     committee on relativity
         experiments, 132
     COBE satellite, 132, 138, 173
     Deep Space Network, 214
     Mars Observer, 214
National Astronomical Observatory
     (Japan), 187
National Bureau of Standards, 217, 218
National Radio Astronomy
     Observatory, 71
National Research Council, 201
National Science Foundation (NSF), 3,
     75, 80, 111, 133, 195
     Binary Black Hole Grand Challenge,
         201–3
     LIGO funding, 138–40, 141, 143,
         146, 154–55, 189
Nautilus, 108, 109
Neodymium YAG, 161
Neptune, discovery, 17
Neutrino detectors, 168–69, 189, 192
Neutron stars, 3, 6, 196. *See also* Pulsars
     audio frequency signal, 204, 207
     binary, 8, 75, 81–82, 84–85, 114,
         166, 173, 192, 203–5, 222, 223
     black hole merger, 171
     boiling, 207
     collision, 84, 96, 168, 171, 203–5

deformations, 207
electromagnetic wave emissions,
     73–74
defined, 9
formation of, 7, 72–73, 206, 207
gravitational field, 53, 203
gravitational waves, 82, 93, 96,
     98–99, 108, 166, 171, 173, 192,
     197, 203
magnetic field, 73
mass, 85, 171
size, 85
"sloshing," 207–8
spin, 73–74, 136, 207
white dwarf stars paired with, 85
Newton, Isaac, 4, 13–17, 89
Newton's laws
     absolute rest frame, 25
     failure of, 17
     of gravity, 5, 13–14, 16, 17, 41–42,
     45, 48
     of motion, 15
     space concepts in, 15, 16–17, 23
     time concepts, 15, 23, 32
NIKHEF (Nuclear Physics and High
     Energy Physics Institute), 109
Niobe, 108
Nobel Prizes, 28, 63, 64, 75, 86, 91, 188
Noise/interference
     for bar detectors, 97–98, 101–4, 107,
     109
     for laser interferometers, 126, 127,
     130, 141, 181
     and interpretation of response,
     102–3, 113
     and isolation and suspension
     system, 93, 179
     LIGO, 152–53, 154, 157, 158,
     159–60, 161, 163, 164, 169, 170,
     172
     shot noise, 161, 163, 168, 182
     for space-based detectors, 215–16
     supercooled bars and, 98, 106–7,
     108, 125, 129
     thermal noise, 97, 156, 157, 163,
     170, 172, 187, 229

Nordtvedt, Kenneth, 63
Norton, John, 41
Nuclear fission, general theory of,
        57–58, 60

O

Occulting disk, 66
Ocean waves, interferometer
        interference from, 154
OMEGA, 219
One, Two, Three, Infinity, 194
Oppenheimer, J. Robert, 59–60, 61, 90
Ørsted, Hans Christian, 25–26
Ostriker, Jeremiah, 211
Oxford University, 129

P

Palomar telescope, 141, 144, 174
Parallax measurements, 20
Parallel axiom, 18, 21
Particle accelerators, 118, 141, 160, 205,
        228
Pennsylvania State University (Penn
        State), 169, 192
Penzias, Arno, 63
Perihelion, 42, 44, 65
Photons, 30, 161
Piezoelectric crystal receivers, 92
Piezoelectric transducers, 104
Planetary probes, 6
Plutonium, 149
Poincaré, Henri, 28, 31, 68n
Portegeis Zwart, Simon, 200
Porter, Cole, 49
Pound, Robert, 54, 58, 129, 130
Press, William, 99, 218
Principia, 13, 15
Princeton Applied Research, 65
Princeton University, 37, 57–58, 62, 63,
        85, 121, 131, 136, 167, 170, 195,
        196, 217
Palmer Physical Laboratory, 59

Plasma Physics Laboratory, 85
Principe expedition, 48–49
Prokhorov, Aleksandr, 91–92
Prussian Academy of Sciences, 39, 43
Pulsars, 5, 61–62, 192. *See also* Neutron
        stars
    binary, 74–75, 78–86, 181
    computer applications, 76–78, 86
    dispersion range, 77
    fastest, 79
    frequency speed, 71, 77, 79
    gamma-ray bursts, 114
    general relativity and, 82–86
    gravitational waves, 82–84, 95–96,
        109, 166, 168, 177, 181
    mass, 75
    Milky Way, 74, 78
    period measurements, 80
    PSR 1913+16, 78–80, 82, 83, 84–85
    radio pulse detection, 71–72, 74, 209
    signal characteristics, 153
    source of, 206
    in Tucanae, 47, 109
Pustovoit, V. I., 124
Pythagorean theorem, 191
Pythagoreans, 191–92

Q

Quality factor, 156, 170
Quantum mechanics, 30, 112–13
Quasars, 6, 9, 50, 57, 70, 192, 195, 209,
        224, 230

R

Raab, Fred, 150, 151, 227
Radar, 117
    signal delays, 51, 62
Radio galaxies, 74, 225
Radio telescopes
    arrays, 6, 8–9, 50, 70, 76
    largest, 76
    millimeter-wave array, 140, 146

pulsar detection, 71, 74, 209
submillimeter, 155
student-built, 74, 75–76
Radio waves. *See also* Pulsars, 27, 50, 96
Rebka, Glen, 54, 58
Redshift, 54–55, 58, 121, 129
Reductio ad absurdum technique, 18
Relativity. *See also* General theory of
    relativity; Special theory of
    relativity
    measurement, 34
Research Electro-Optics, 156
Richstone, Douglas, 224
Riemann, Bernhard, 20–22
Riemannian curvature, 20–22, 41, 45
Riemannian geometry, 41, 45
Riemannian manifold, 35, 41
Römer, Ole, 24, 25
Royal Astronomical Society and the
    Royal Society of London, 49

S

Saccheri, Girolamo, 18
Sagan, Carl, 194
SAGITTARIUS, 219
Sagittarius constellation, 79, 96
Sandeman, John, 187–88
Sanders, Gary, 167, 190
Sanduleak −69° 202, 6–7
Sapphire mirrors, 170, 188
Saulson, Peter, 70, 87, 111, 123n, 138,
    144, 170–72
Schilling, Roland, 180, 181–82, 183
Schutz, Bernard, 185, 219
Schwarzschild, Karl, 60
Schweikart, Ferdinand, 19
Sensitivity
    of laser interferometers, 124,
        127–28, 132, 135, 136, 181,
        182–83, 186
    of LIGO, 142, 152, 153, 161, 169,
        186
    of resonant bar antennas, 105, 107,
        109–10, 114

signal recycling and, 186
size of instrument and, 186
Seventh International Conference on
    General Relativity and
    Gravitation, 104
Sfera Project, 110
Shapiro, Irwin, 50–51, 62
Shoemaker, David, 143
Shot noise, 161, 163, 168, 182
Silicon dioxide-tantalum pentoxide
    coating, 156
Silk, Joseph, 142
Singularities, 60, 196, 210
Snider, Joseph, 54
Snyder, Hartland, 59–60
Sobral expedition, 48, 49
Solar eclipse experiments, 48–49, 50,
    51–52
Solar wind, 214, 216
Soldner, J., 48n
Sommerfeld, Arnold, 43
Space. *See also* Void
    absolute, 16, 23, 32
    contraction of, 32
    esoteric concepts, 11–13
    Euclidean, 18, 21, 22
    metrical field, 22, 45
    negative curvature, 19, 21
    Newton's concepts, 15, 16–17
    positive curvature, 20–21
    theological concepts, 12, 16–17
Space-based gravity wave detectors
    atomic clocks in, 215
    chances for success, 224, 226
    events detectable by, 213, 222–24
    frequency, 215, 216, 222, 225
    LAGOS, 219
    laser interferometers, 216, 219–21
    LISA, 219–224, 225–26
    noise/interference, 215–16, 218, 223
    OMEGA, 219–20
    orbit, 218, 219–20
    principle, 214, 215
    spacecraft as, 214–15, 216, 218
    sensitivity, 216
    test masses, 220, 222

Spaceborne clocks, 6, 215
Space-time
  black holes and, 199, 200–1
  concept, 4–5, 34–35
  curvature, 41, 44–46, 58
  dragging, 55–57, 82, 204, 222
  electromagnetic waves and, 193
  at event horizon, 60
  expansion of, 8
  general relativity and, 4, 44
  gravity and, 41, 43–46, 55–57
  gyroscope experiment, 56–57
  ripples. *See* Gravitational waves
  rotation, 55–57
  warping, 45, 48, 51, 52, 55–57, 60,
    122
Special theory of relativity
  equation, 34
  four-dimensional nature, 35
  frame of reference, 29, 31, 32, 33
  gravity and, 38–39
  mathematical arguments, 31
  Minkowski's reinterpretation,
    34–35, 41
  physicists' reactions to, 35, 37
  speed of light, 31, 32, 33
  time, 31–32, 33, 35
Speed of light
  early measurements, 23–24, 26–27
  gravitational waves from matter
    approaching, 68, 192–93
  laser measurements, 98
  and mass, 34
  special theory of relativity, 31, 32,
    33
  through matter, 32n
  in vacuum, 32, 72, 123n
Spherical bar detectors, 109–10,
  176–77
Stanford Linear Accelerator, 167–68
Stanford University, 95, 97, 107, 129,
  169, 183
Stars
  collapse, 60
  collisions, 68
  gravitational radiation, 68

light deflection, 43, 45, 48–50, 51–52
parallax measurements on, 20
white dwarf, 85
Stebbins, R. Tucker, 97–98, 218–19, 220,
  222
Strain, 68, 69, 87–88, 98–99, 100–1,
  131, 135, 137, 161, 169, 184,
  216
Strong Equivalence Principle, 217
Sun (Sol)
  oblateness measurements, 65, 66
Superconducting instrumentation,
  107
Superconducting Super Collider, 146,
  147, 155, 164, 167
Supercooled bars, 98, 106–7, 125, 129,
  180
Supernovas, 3
  gamma-ray bursts, 114
  gravitational waves, 68, 69, 88, 90,
    95, 96, 105, 108, 110, 153, 166,
    171, 177, 178, 192–93, 198,
    205–7
  Milky Way, 105, 153
  neutron stars, 72–73, 79, 205–6
  1987A, 110, 136, 154, 206, 207
  Sanduleak–69° 202, 6–7
  signal characteristics, 153, 206
  strain, 223
  white dwarf collisions, 226
Signal recycling, 186, 188
Syracuse University, 70, 169, 171

T

TAMA 300, 187
Taylor, Joseph, 70, 71–72, 74–75, 79,
  80–81, 82–85, 86, 87, 181
Telescopes, 8
Texas Symposia on Relativistic
  Astrophysics, 84, 100
Thermal noise, 97, 156, 157, 163, 170,
  172, 187, 229
Thiokol Chemical Company, 195
Thirring, Hans, 55

Thorne, Kip
   bets, 210
   black hole/neutron star research,
      63, 195–98, 199, 201, 203,
      205
   and Braginsky, 131, 133
   cosmic strings, 195
   and Drever, 133
   and experimentalists, 196–97
   general relativity work, 62, 63, 131,
      195, 196–97
   on Gertsenshtein, 124
   Gravitation, 6, 82, 132, 142, 197
   gravitational wave research, 82, 95,
      99, 131–33, 177, 193, 194, 198
   laser interferometry research,
      133
   LIGO construction, 138–39,
      140, 141, 169, 193, 194
   personal background, 194–95
   rocket engine design, 195
   space-based detectors, 218
   and Weber, 95, 99, 131–32
   and Weiss, 132, 138, 193
   wormhole theory, 193–94
Thunderstorms, 154
Tidal actuators, 154
Tilav, Serap, 168–69
Time. *See also* Space-time
   absolute, 16, 23
   contraction of, 32, 33, 53
   in gravitational field, 52–53
   Newton's universal clock, 15–16,
      32
   in special theory of relativity,
      31–32, 33, 35
Tomorrow Never Dies, 55
Townes, Charles, 91
Treatise on Electricity and Magnetism,
   26
Trimble, Virginia, 112
Tucanae 47, 109
Twente University, 109
Tyson, J. Anthony, 100–1, 104, 105,
   106–7, 113, 141–42

U

Ulysses, 214
Universal constant, 32
Universal expansion, 58, 64, 208,
   226–27
Universe size, 204
University of
   Amsterdam, 109
   Berlin, 39, 42
   California at Irvine, 111
   Chicago, 100, 140, 196
   Colorado, 169, 217
   Delaware, 168
   Florida, 169
   Glasgow, 95, 104, 128, 129, 131, 133,
      134, 183, 186
   Göttingen, 19, 20
   Hannover, 184
   Leiden, 90, 109
   Maryland, 10, 87, 89, 91, 93, 94, 98,
      104, 106, 111
   Massachusetts at Amherst, 74, 76,
      81
   Michigan, 169
   Oregon, 169
   Padua, 8
   Rochester, 96–97, 100, 103, 104
   Rome, 106
   Texas at Austin, 202
   Washington in Seattle, 64
   Wisconsin, 168, 169
   Zürich, 39, 42
Uranus, orbital motions, 17
U.S. Air Force, 205
U.S. Department of Defense, 136
U.S. Department of Energy, 149, 155
U.S. Navy, Bureau of Ships, 91

V

Vacuum systems, 117, 134, 151, 158–60,
   185–86

Vega, 71
Vela satellites, 205
Venus, radar signals to, 51
Vessot, Robert, 54, 121
Viking landers, 51
Virgo cluster, 105, 171, 178, 200, 206
VIRGO detector, 176–80, 186, 222
Vogt, Rochus (Robbie), 140–41, 142–43, 145–46, 147
Void
    continuous three-dimensional, 15
    pneuma apeiron concept, 12
von Eötvös, Baron Roland, 64

W

Walther, Herbert, 183
Washington University at St. Louis, 63
Weber, Joseph, 10, 87, 193
    contribution to gravity wave research, 180, 195
    correlations with other data, 106–7, 110–11, 114–15, 180
    criticisms of, 98–104, 105, 132, 138, 189
    defense of findings, 101, 102, 103–4, 106–7, 112–13
    expertise, 102
    gravitational wave antennas, 89, 90, 92, 93–94, 102–3, 154, 195, 203
    gravitational wave observations, 94–96, 97, 180
    laser interferometer, 124–25
    maser concept, 91, 112
    observatory, 113–14
    personal background, 90–91, 111–12
    quantum mechanics explanation, 112–13
    relativity research, 88–89, 91–92

Weisberg, Joel, 83
Weiss, Rainer, 68, 85
    atomic clocks, 118, 121
    Dicke and, 65, 121, 126
    on Einstein, 4
    first-detection ground rules, 188–90
    as graduate student, 121
    gravimeters, 121
    gravitational constant measurement, 122
    laser interferometry research, 122–23, 125, 131, 133–34, 136–37, 181, 218
    LIGO construction, 117, 119, 137–40, 141, 159–60, 173, 227–28
    microwave background measurements, 132, 170, 173
    at MIT, 117–18, 120, 121–22, 131, 136
    NASA committees, 132, 216
    personal background, 119–20
    personality, 117, 140
    space-based detectors, 216
    on theorists vs. experimentalists, 118–19, 193
    Thorne and, 132, 138, 193
    as undergraduate, 118
Western Reserve University, 27
Weyl, Hermann, 22
Wheeler, John Archibald, 6, 57–58, 59–62, 67
    Gravitation, 6, 82, 132, 197
    Hanford Nuclear Reservation, 149
    personal characteristics, 63
    Thorne and, 82, 195, 197
    Weber and, 90, 92, 195
Whitcomb, Stan, 134, 155, 160
White, Martin, 208
Will, Clifford, 5, 55, 63–64, 213
Wilson, Robert, 63
Wizard of Oz, The, 200
Wormholes, 193–94

X

X-ray astronomy, 9, 192, 196, 197, 199, 210

Y

Yale University, 129
Yamamoto, Hiro, 164

Z

Zacharias, Jerrold, 120–21
Zapolsky, Harry, 133
Zipoy, David, 93
Zucker, Michael, 154, 161–62, 163
Zwicky, Fritz, 52, 72